Plant Bioactives And Engineered Nano-Carriers In Protein Fibrillation

by

Aalok Basu

Preface

Degenerative disorders are one of the leading causes of deaths worldwide. Different degenerative diseases are increasingly recognised as the major factor behind life-long morbidity and mortalities. As the global population is growing and aging, the incidences of degenerative disorders are escalating sharply. This presents a great challenge in population health management, cost, and governmental concerns. Existing healthcare systems are facing an increasing demand for proper therapeutic support and rehabilitation services for the affected population, particularly so in the low-income and middle-income countries. Global Burden of Diseases (GBD) data have expressed the need for newer knowledge to formulate strategies towards efficient mitigation of degenerative disorders. Current researches in medicines have consequently focused on backward integration of progressive events in the molecular and cellular levels, culminating in the fundamentals of pathophysiology. More than 37 degenerative disorders including Alzheimer's disease, liver fibrosis, Parkinson's disease and type 2 diabetes are reportedly associated with misfolding of certain amyloidogenic proteins and their subsequent fibrillation. Understanding the mechanism of protein fibrillation and inhibition of the protein misfolding through stabilization of their three dimensional structures are currently considered as tools for designing appropriate therapeutics to alleviate different degenerative disorders.

Some bioactives isolable from indigenous plant sources have been established as promising compounds and as alternative therapeutics in a multitude of degenerative diseases. However, their actions at near molecular levels are still very poorly understood. Amidst different chemical classes, the terpenoids and flavonoid type compounds are presently explored as lead molecules in drug discovery research and to boast efficacy at specific bio-molecular interfaces. These compounds in purified forms, however, pose a challenge in systemic delivery owing to their solubility constraints and rapid metabolic clearance. Principles of nanoparticle engineering were therefore conceived in order to address some such constraints and harvest the therapeutics potentials of these compounds. Mitigation of a range of degenerative disorders now appears a possibility furnished through appropriate interventions in protein fibrillation processes. Functional nanoparticles associate distinctive advantages including Brownian motion in biological fluids, rapid diffusion and directed protein interactions. Enhanced pharmacodynamics due to engineered nanoparticles often presents yet another advantage in the nano-bio interfaces.

This work focuses on exploration of the effects of a diterpene lactone andrographolide, a flavonol quercetin and a flavonolignan silybin in protein fibrillation interfaces. Andrographolide is a typical diterpenoid used in treatment of diverse medical conditions, including a few degenerative disorders. Quercetin is one known free-radical scavenger and has showcased potential therapeutic activity in oxidative stress related ailments, such as drug induced hepatotoxicity. Silybin is another bioactive identified as a therapeutic for hepatic regeneration and as nutraceutical. Extensive wet lab and *in silico* studies were designed for understanding protein fibrillation effects and mechanisms. Nano-therapeutics were also considered proficient against fibrillating plasma proteins. This is because of nanoparticles' rapid diffusion, unique surface chemistry, and Brownian motion. Plant bioactives tethered gold nanoparticles were conceived for exploring possible repression of protein fibrillation. The behaviour of nanoparticles against fibrillating protein required in depth study for a thorough understanding.

The course of this research work is presented in different chapters:

The first chapter introduces the concept of protein misfolding, and the consequent degenerative disorders. Intervening in progression of protein fibrillation is considered as one trending approach for control. The chapter also presents a review of literature with emphasis on selective plant bioactives effects, and some state-of-the-art nano-carriers developments.

The second chapter highlights the fundamental background and presents the major objectives of the research work. Serum albumin was used as the model protein throughout the studies. Standardized fibrillation techniques were adopted for *in vitro* analysis. This part also consolidates the various steps taken to address the purpose.

The third chapter explores the efficacy of andrographolide in protein fibrillation. The principles of protein-bioactive interactions were applied in order to comprehend the effects of the diterpenoid labdane molecule in protein fibrillation interfaces.

The fourth chapter concentrates on the effects of quercetin at various levels as one inhibitor of protein fibrillation. The flavonoid was also tested as a disruptor of preformed fibrils.

The fifth chapter deals in the study of silybin as protein fibrillation inhibitor. The protein-bioactive interactions were understood in biomimetic conditions, and the effects of the flavonolignan molecule were investigated.

The sixth chapter is focused on gold nano-carriers surface tethering with various bioactives. Gold nanoparticles outspread a stratagem to deter protein fibrillation. Nano-chemistry of synthesized gold nanoparticles was explored and the effects on protein fibrillation studied thoroughly. Plant biomolecules stabilized gold nanoparticles were perceived as safer tools against protein fibrillation.

The seventh chapter is concerned with exploring the protein corona formation around biopolymer nano-carriers. Nanoparticle protein corona is often considered as a resultant of insoluble proteins deposition. Surface chemistry of biopolymer in nanoscale was considered as a strategy to influence protein corona effects.

The eighth chapter consolidates the efficacy of nano-encapsulated quercetin in hepatotoxic conditions. Paracetamol overdose is known to induce endoplasmic reticulum stress in liver, leading to tissue injury, and subsequent fibrosis. Biopolymer based nanomedicines were further developed to alleviate similar experimental conditions.

The ninth chapter summarizes the findings of the entire research work and presents a perspective. The references were organized using 'Mendeley Desktop', version 1.19.4.

The annexure part presents the list of various publications, presentations and awards received during the tenure of dissertation work.

High global burden of degenerative disorders have invoked an urgent need for newer strategies for disease control. Inhibitory effects on amyloidogenic protein fibrillation were considered as one strategy. Proteins fibrillation and corona formations in mimicking conditions were studied in depth and the potentials of engineered nano-carriers stabilized with selected plant bioactives were evaluated. Five research articles related to the work title have been published so far in journals of repute. Five presentations pertaining to the work chapters were made in different conferences nation-wide and one work has been credited with the best presentation award.

Contents

Page no.

CHAPTER **1**

Introduction

1. Introduction

Proteins are exquisitely arranged molecules which often carry out various well-defined biological functions. Proteins in human, serve a variety of purposes. Some are structural proteins like keratin of hair, while some are functional like the insulin or human serum albumin, and others are storage devices such as the casein in milk or ferritin in liver [1,2]. Proteins bio-functions are mostly determined by their three dimensional macromolecular structures, spatial arrangement of functional groups and interplay of various non-covalent interactions. Physiological stress conditions such as oxidative imbalance enhanced cellular metabolism or dominant negative mutation, may initiate biochemical cascade in which the protein molecules tend to self-associate into amorphous aggregates. That gives rise to extracellular fibrillar structures which are toxic in nature and are difficult to be metabolized. Similar molecular disposition brings about cell death and/or persistent and serious pathological abnormalities [3]. The process is termed as amyloidogenesis. Protein misfolding and fibrillation are underlying causes in a number of degenerative conditions like Alzheimer's disease, Parkinsonism, type 2 diabetes, atherosclerosis and others [4]. Therapeutic strategies to mitigate or prevent amyloidogenic diseases currently include stabilization of native protein structures and increased clearance of already misfolded protein aggregates [5,6]. Some bioactives from plant sources expressed a propensity to alleviate in similar degenerative conditions but that effect has remained very implicit so far. Understanding the molecular events leading to protein fibrillation and its rightful intervention are urgent for taking a control over rapid spread of different degenerative diseases. The subject area is very demanding and lies at the interfaces of biophysics, biochemistry, medical chemistry, pharmacy, pharmaceutics and delivery design. The introduction part presents a review through major specialization areas in line with the work in focus. Various literatures were studied, consolidated and aligned in the references section using Mendeley desktop, version 1.19.4.

1.1. Protein fibrillation: misfolding, aggregation and amyloidosis

Proteins constitute the structural and motor components of cells and they are also responsible for a multitude of coordinated functions in bio-system [7,8]. Similar biomolecules are synthesized based on the information embedded in the cellular DNA. Multiple amino acids with unique side chains are arranged in a linear sequence to form the primary structure of a protein [9]. The side chain chemistry is critical to the conformation of the proteins, since the functional groups can interact with one another through hydrophobic, hydrogen, ionic or other covalent interactions. These interactions and location of amino acids in the linear sequence cause a protein to fold and assume a specific, compact three dimensional structure. As the protein folds, it tests different conformations before reaching its unique, thermodynamically stable form.

Water molecules extend precise roles in determining the protein stability, dynamics, 3D-structure, and functions in the bio-environment [10]. The hydration forces are particularly responsible for packing and stabilization of the protein structure. Protein folding is largely depended upon various non-covalent interactions depend upon water positioning. Water is fundamental in protein folding mainly because of its role in defining hydrophobic attractions responsible for the kinetically dependent gluing of hydrophobic residues [11].

The folded protein may further be stabilized by hundreds of non-covalent bonds between the amino acid residues [12,13]. Studies revealed that various factors such as molecular chaperones and folding enzymes present in the cellular environment contribute immensely to the folding process [14]. Chaperones can attach to the partially folded proteins and guide to achieve the correct form by harvesting molecular energy from the energy rich phosphates. Some enzymes such as protein disulphide isomerase or the peptide propyl isomerase also help in the catalysis of protein folding [15].

Protein folding also occurs in a spontaneous manner based on the inherited molecular information, environmental stress and related factors. Somatic mutations, translational and post-translational errors, malfunction of chaperones, structural modifications due to environmental stress are some of the many factors that may cause a protein to undergo an erratic folding [3]. This phenomenon termed as protein misfolding results in formation of non-functional protein structures. Eukaryotic cells comprise of efficient quality control machineries to check the occurrences of protein misfolding within the cell and degrade the misfolded proteins [16]. Apart from the molecular chaperones and enzymes, the quality control machineries involve endosomal- lysosomal system, ubiquitin- proteasome system, unfolded protein response, and heat shock proteins [17,18]. The protein homeostasis or proteostasis system initially detects the misfolded proteins. In early stages, refolding of protein is attempted by the molecular chaperones. If refolding process is unsuccessful, the ubiquitin- proteasome system and lysosomes degrade the misfolded proteins to prevent their self -aggregation. Heat shock proteins detect the exposed hydrophobic groups of misfolded proteins and bind with them to evade self-aggregation. The misfolded proteins -heat shock protein complexes are recognized and ubiquitinated by ubiquitin ligases. On the other hand, endosomal- lysosomal systems involve formation of endosomes containing ubiquitinated proteins. Endosomes further can fuse with the lysosomes to facilitate clearance from the cells [15].

Although the occurrence of protein misfolding is monitored by the cellular quality control systems, many proteins fail to fold accurately and become misfolded. It is well known that monomers of misfolded proteins facilitate specific intermolecular interactions which results in self-association of the monomers to form oligomers [19,20]. Formation of small, reversible oligomers progressively leads to accumulation of larger aggregates, and transforming into toxic, irreversible structures. These structures are mostly characterized by the presence of β folds which are stacked perpendicular to the aggregation axis. Hydrogen bonds between the polypeptide carbonyl oxygen and amide group drive the formation of the β folds. Hydrophobic interactions between bulky side chains, and salt bridges between acidic and basic groups further stabilize the integrity of the aggregates. Protein aggregation through self-association of misfolded proteins initiates the formation of proto-fibrils, which matures into toxic amyloid fibrils. Amyloid fibrils are formed through twisted linking of 2-6 protofibrils and have diameter of 4-13 nm [21]. Deposition of these amyloids in intracellular or extracellular regions, termed as amyloidosis and subsequently leads to serious degenerative conditions. The composition of amyloid fibrils varies with different precursor proteins, though the width of fibres are often similar.

Earlier, amyloid fibrils in pathogenic conditions were difficult to study due to large structure, poor solubility and quazi-crystalline nature. However, the understanding of protein aggregation under *in vitro* condition is now increased through application of emerging technologies. The structural hallmark of the protein fibrils is the presence of cross β sheets, which are responsible for tinctorial capacity of the structures. Extensive studies through solid state NMR and XRD revealed that the cross β sheets are arranged in parallel or anti-parallel orientation [22,23]. The morphology of fibrils formed under *in vitro* conditions depends upon conditions such as protein concentration, pH, temperature, buffer composition and presence of additives.

Different pathways or mechanisms guide the fibrillation of proteins. The mechanisms of protein fibrillation have been explored in order to understand the prevalence of amyloid-related degenerative diseases. Some of the well understood mechanisms have been described in the following section:

(i) Reversible association of protein monomers

Native protein monomers can self-associate and extend intermolecular interactions to form small, reversible oligomers. Further transformation into larger oligomers results at higher protein concentrations and these aggregates progressively become irreversible through formation of covalent bonds. This mechanism also depicts that union of protein molecules through end-by-end or side-by-side combinations can lead to aggregation, and protein misfolding is not an obligatory condition [24–26].

(ii) Association of conformationally altered protein monomers

Some proteins do not have any tendency to self-associate reversibly under normal conditions. Stressed conditions such as high temperature, low pH, and organic solvents causes the protein monomers to lose their native conformations. These conformationally altered units however are prone to associate, and form insoluble aggregates through specific interactions. It is apparent that conditions or compounds which stabilize the native structure of the proteins may prove helpful in designing suitable approaches to inhibit protein aggregation [27–30].

(iii) Nucleation based aggregation

Nucleation controlled aggregation is the most widely studied mechanism which involves the formation of visible aggregates. The process begins with a lag phase without any visible protein aggregates. After a certain time period, several high-energy species of critical size appear that are structurally rich in β sheets. It should be noted that monomers with a tendency to form oligomers does not necessarily favour aggregation unless the critical size of the aggregates is achieved. The aggregates then grow larger in size through continuous addition of monomers [31,32].

(iv) Surface induced aggregation

Surface-monomers interactions occur when the native protein bind to the surface of a container. Such interactions may be both hydrophobic and electrostatic and cause conformational changes to the monomers. These conformational alterations increase the tendency of the monomers to aggregate at the surface or at the bulk of the solution [32–34].

Though these models of protein fibrillation improve the facile understanding of the molecular causes of amyloid related diseases, it is obvious that the conditions within biological systems are far more complex and several kinds of aggregates do co-exist.

1.2. Protein fibrillation in various diseases

Till now, about thirty seven proteins have been identified that form amyloid deposits in human physiological system (Table 1.1). Seven of the amyloidogenic proteins are associated with neurodegenerative disorders such Parkinsonism and Alzheimer's diseases, while the rest are mostly responsible for non-neuropathies such as systemic amyloidosis, diabetes and various carcinomas. The amyloidogenic proteins can be structurally categorized into two categories: One which have a well-defined three-dimensional structure, and another which are intrinsically disordered with low thermodynamic stability [8,35]. These proteins often deposits in the extracellular spaces, though few of them are reported to form intracellular inclusions. More recent studies provide ample evidences of the presence of both extracellular and intracellular aggregates in most of the amyloid related degenerative conditions [36–38]. Several workers have reported that the amyloid aggregates can move in between intracellular and extracellular compartments and contribute to the disease pathology, as in case of Alzheimer's disease [39].

Occurrences of Alzheimer's disease have been linked to extracellular deposition of β amyloid peptide in brain parenchymal tissue as well as in the brain neuron. The β amyloid peptide is one of widely studied, intrinsically disordered protein arising from endoproteolysis of amyloid precursor protein [40]. Another intrinsically disordered protein, serum amyloid A peptide, is a product of chronic inflammation pathways and causes AA amyloidosis upon aggregation [41,42]. The fibrillation propensity of the intrinsically disordered proteins is governed by their concentration and post translational modifications. Other proteins of similar characteristics include tau (τ) that aggregates to cause frontotemporal dementia [43], and α synuclein whose aggregation is associated with Parkinsonism [44,45].

Table 1.1: Some amyloidogenic proteins and associated diseases

Protein or peptide name	Related diseases	Type of disease
α synuclein	Parkinson disease	Neuropathic
β amyloid peptide	Alzheimer's disease	Neuropathic
Prion protein	Huntington's disease	Neuropathic
τ protein	Frontotemporal dementia with Parkinsonism	Neuropathic
Huntingtin exon 1	Huntington's disease	Neuropathic
ADan peptide	Familial dementia	Neuropathic
ABri peptide	Familial dementia	Neuropathic
Immunoglobulin (Ig) light chain fragments	Light chain amyloidosis	Systemic
Ig heavy chain fragments	Heavy chain amyloidosis	Systemic
Serum amyloid A protein fragments	AA amyloidosis	Systemic
Transthyretin	Systemic amyloidosis	Systemic
β_2 microglobulin	Systemic amyloidosis	Systemic
Apo-lipoproteins	Amyloidosis	Systemic
Lysozyme	Visceral amyloidosis	Systemic
Fibrinogen fragmens	Renal amyloidosis	Systemic
Cystatin C	Cerebral haemorrhage	Systemic
Islet amyloid polypeptide	Type II diabetes	Localized
Calcitonin	Carcinoma in thyroid	Localized
Atrial natriuretic factor	Atrial amyloidosis	Localized
Prolactin fragments	Pituitary prolactinoma	Localized
Insulin	Amyloidosis in injected area	Localized
Medin	Aortic medial amyloidosis	Localized
Lactoferrin	Gelatinous corneal dystrophy	Localized
Odontogenic ameloblast associated protein	Odontogenic tumours	Localized
Pulmonary surfactant associated protein	Pulmonary alveolar proteinosis	Localized

Leukocyte derived chemotaxin 2	Renal amyloidosis	Systemic
Galectin 7	Lichen amyloidosis	Localized
Corneodesmosin	Hypotrichosis simplex in scalp	Localized
Kerato-epithelin fragments	Corneal dystrophy	Localized
Semenogelin 1	Seminal vesicle amyloidosis	Localized
Proteins S100A8/A9	Prostate cancer	Localized

Studies on well-defined proteins, such as Transthyretin and Immunoglobulin light chain exhibited that any conformational alteration can misassemble the proteins into amyloid fibrils and similar aggregates [46]. The basis of their aggregation is explained by the conformational alteration hypothesis, and the process is mediated by different hydrogen bonds and hydrophobic interactions. On contrary, Prion proteins can form amyloid-like aggregates through nucleation and seeding mechanism and have major pathological roles in transmission of spongiform encephalopathy and Creutzfeldtz- Jakob syndrome [47,48].

A number of genetic, pharmacological and biochemical studies showed that mutations cause certain genes to encode for fibril forming proteins. Mutations in fibril forming proteins have been linked to early onset of Huntington, Creutzfeldtz- Jakob, Alzheimer's, and Parkinson's diseases [49]. Hence, an increasing number of evidences favour the hypothesis that protein fibrillation is a critical step in the diverse degenerative diseases. The β sheet rich protein fibrils localize in the tissues and contribute to cell death. Inhibition of protein fibrillation and/or disruption of the fibrils are envisaged to be the possible therapeutic strategies for their mitigation.

Although there are a number of proteins that can form amyloid fibrils (e.g. insulin, α synuclein, lysozyme, $\beta2$ microglobulin) under *in vitro* conditions, human serum albumin (HSA) is the most widely studied model protein used to understand the molecular effects of various inhibitors or disruptors of fibril formations [50,51]. HSA is a typical α-helical (> 60 %), single chained protein containing 585 amino acids arranged in three homologous domains [52]. Literature have shown that the systematic aggregation of albumin occurs through partial misfolding of its native structure and the process has been conveniently replicated *in vitro* through specific conditions such as high temperature, metal ions, treatment with denaturing chemicals, and lowering of pH [53].

1.3. Plant bioactives against amyloidogenic diseases

Plants constitute an integral part of medicines since the Neanderthal period [54]. Various indigenous systems of medicines such as Ayurveda, Chinese and Unani have records of several plant species to alleviate disease conditions. In India alone, almost 20,000 medicinal herbs have been used by traditional communities to cure different ailments [55]. However, the complexity of plant products limited the scope of chemical modification necessary for optimized therapeutic applications. In the present times, the use of cinchona, morphine, digitalis and introduction of aspirin, a plant based derivative, led researchers to further exploit the diverse wealth of plant resource. Many such compounds derived from plants used in traditional medicine have become an essential part of the health care system. Plant bioactives have proven invaluable as effective tools for understanding the logics of biosynthesis and as sustainable platforms for designing blockbuster drugs [56]. These compounds offer a wide range of structural diversity [57], and consideration of plant products as alternative sources of newer therapeutics have been achieved through modern means of separation, screening, structure elucidation and synthesis.

The plant bioactive compounds exhibiting anti-amyloidogenic capacities in neurodegenerative conditions belong to a wide range of phytochemical groups: polyphenols, terpenoids, carotenoids and alkaloids [58]. Different classes of polyphenols including stilbenes, lignans and flavonoids are known for their antioxidant capacity through quenching of endogenous oxygen species and inhibit the progression of amyloid related diseases [59]. Some polyphenols such as curcumin, epigallocatechin-3-gallate (EGCG) and resveratrol have made up to clinical trials for control of Alzheimer's disease [60]. These molecules also involve pleotropic mechanisms that reduce β amyloid production, and eventually slow disease progression [61]. Scopus database showed that flavonoids such as quercetin, kaempferol, rutin, catechin, apigenin and several anthocyanins are often reported for treatment in progressive dementia. Alkaloids such as berberine, anatabine, oridonin, and galantamine abrogate the inflammatory progression of the disease through selective inhibition of NF-κB acetylation [62]. Compounds such as ginsenoside [63], curcumin [64], green tea polyphenols [65] and resveratrol [66] have exhibited protective actions on dopaminergic neurons from certain neurotoxins in Parkinson's disease. They limit the loss of dopamine, inhibit microglial activation, reduce the levels of pro-inflammatory factors, and modulate the protein expression of apoptosis-related factors [67]. Studies on gallic acid and polyphenols revealed that these antioxidants prolong and increase locomotor activity in experimental conditions [68].

Many of the compounds have been proven equally efficacious in type 2 diabetes as well. Flavonoids, such as naringin, exhibit potential control on serum dipeptidyl peptidase-4 levels (DPP-4), and restrict random blood glucose levels. Naringin also causes hepatic glycolysis and decrease gluconeogenesis in liver to mitigate hyperglycaemia [69]. Baicalein has been reported to improve the islet β cell survival of diabetic mice [70], inhibit oxidative stress and impede inflammatory responses [71]. Another well-known flavonoid, quercetin, reduces liver inflammation and controls glucose homeostasis through stimulation of GLUT4 translocation in voluntary muscles [72]. Quercetin also ameliorates diabetic neuropathy by limiting the expression of connective tissue growth factor and transforming growth factor $\beta1$ in diabetic rats [73]. Other flavonoids capable of alleviating hyperglycaemic conditions include myricetin [74], kaempferol [75], apigenin and icariin [76]. Plant terpenoids also play significant roles as anti-diabetics in indigenous systems of medicine. Genipin can relieve oxidative stress and reduce steatosis to attenuate hepatic insulin resistance under experimental conditions [77]. Oleanolic acid is reported to reduce blood glucose, improve insulin tolerance and inhibit hepatic gluconeogenesis [78]. Andrographolide is another terpenoid which markedly diminished production of oxygen species and mitigated diabetes neuropathy through inhibition of renal inflammation and fibrosis [79]. Some alkaloids including berberine [80], capsaicin [81], piperine [82], theobromine [83] and colchine [84] have been used both traditional and modern medicines for discovery of hyperglycaemic drugs. Among them, berberine has proven to enhance insulin secretion, and reduce blood glucose level via activation of different pathways.

Several of plant bioactives furthermore have demonstrated potential effects in secondary amyloidosis in other tissues such as liver [85], kidney and spinal cord [86]. Generation of reactive oxygen species (ROS) and chronic infections in liver contribute to liver fibrosis [87], and eventually lead to cancer [88]. Glycyrrhizin, a triterpene obtained from liquorice, can inhibit hepatocyte apoptosis by subduing caspase-3 expression and prevent the release of cytochrome C from mitochondria [89]. A flavonolignan, silymarin, is a well-studied compound for liver related conditions in both clinical and experimental settings [90]. It reduces ROS production and lipid peroxidation, and inhibits the binding of various toxins to hepatocyte cell receptors [91]. Apart from exerting a membrane stabilizing effect, the molecule promotes regeneration of hepatocytes, controls inflammatory responses and inhibits liver fibrosis [92]. Drug or heavy metal induced toxicity of liver has been found to be alleviated through regular treatment of quercetin. Quercetin administration cause significant reduction of inflammatory biomarkers and protect the tissues from oxidative stress and mitochondrial damage [93,94]. Naringin is another candidate which exerts protective and anti-fibrogenic action on both drug and heavy metal induced hepatotoxicity [95,96]. These flavonoids and chemically related compounds may therefore help prevent development of hepatic fibrosis in oxidative stress conditions. Other compounds such as rhein protect both liver and kidney tissue from oxidative damage in paracetamol induced hepatotoxicity [97]. The pharmacological mechanism includes antioxidative and anti-inflammatory property and reduction in activation of hepatic stellate cells [98]. Various constituents from traditional medicine, namely curcumin [99], ellagic acid [100], matrine [101], artemisinin [102] are also being presently explored as lead compounds for mitigation of liver diseases.

Much of the works on "mechanism of action" of these plant bioactives are preliminarily based on pharmacological and *in vitro* biochemical indications. Information regarding molecular targets of plant bioactives though increased in recent years, is yet but sparse. Some of these compounds not only demonstrated multi-target effects, but also complement with the existing therapeutic regime for synergistic outcomes [103]. It is therefore envisaged that a detailed study of bioactive molecule interactions in protein fibrillation interfaces would help to gain some insights of their mechanism in amyloidogenic diseases.

1.4. Nanoparticles as biomolecular carriers

The use of nanotechnology in delivery of plant bioactives is a domain of immense research interest. Nanoparticles of both polymeric and metallic origins are currently being explored as tools for enhancing the bioavailability, thereby the effectiveness of various biomolecules. Besides, nanoparticles possess certain advantages: (i) They can penetrate the smallest of capillaries owing to their nanometric size (less than ~100nm) and escape rapid clearance by the phagocytes; (ii) they can pass through cells and tissue gap to reach a specific organ; (iii) they exhibit controlled release capacity due to pH change, biodegradability, and thermo-sensitivity of the material [104]. The nanoparticle-based platforms have thus emerged as suitable carriers for addressing the pharmacokinetic constraints associated with the conventional formulations, thus ensuring site specific delivery of payload, lesser side effects and higher therapeutic outcome [105,106].

Large number of the well-known bioactives such as curcumin [107], quercetin [108], resveratrol [109], and andrographolide [110], are highly lipophilic and exhibit low bioavailability. Administration of a large quantity of the bioactives is required to produce the optimum therapeutic effect, however increasing the chances of acute toxicity [111]. Encapsulation of these molecules in colloidal carriers improve their water solubility and therapeutic efficacy. Experiments to enhance of their bioavailability through nanoparticulation have been routinely carried out since the past few decades. One pharmacokinetic study demonstrated that relative oral bioavailability of quercetin solid lipid nanoparticles in comparison to crude quercetin suspension was 571.4 % [112]. Orally administered curcumin in nano-forms showed a marked increase in systemic bioavailability in rat model, in comparison to its crude suspension. The AUC values of liposome-encapsulated curcumin nanoparticles were 4.96 times higher than curcumin [113]. Semi-synthetic polymers are also being experimented for improving the absorption of plant bioactives. Hydroxypropyl methylcellulose was applied for encapsulating icaritin through antisolvent-precipitation method, and the pharmacokinetic parameters were studied in rat model. The nanosystem exhibited faster dissolution and higher absorption following oral administration [114]. Several workers have attempted to increase the bioavailability of various other plant bioactives including silymarin [115], kaempferol [116], stevioside [117], aureusidin [118], tea polyphenols [119], resveratrol [120] and daidzein [121] through nanoparticulation using different biopolymers.

A major benefit offered by nanoparticles in delivery of their payload is tissue or organ specific target-ability. Delivery of plant bioactives to specific points reduces the chances of toxic effects of the compounds. Such delivery strategies are generally achieved through functionalization of the nano-carrier surface with various proteins, peptides or small molecules [122]. Transferrin and folic acid are prime examples of molecular tethers that mediate the delivery of nano-therapeutics in the brain. Targeting of payload for subcellular organelle delivery is now achievable which would further extend the scope of drug efficacy [123]. These target-able medicines would therefore offer effective control over amyloid-related neurodegenerative disorders including Alzheimer's disease, Parkinsonism and others [124]. Nanoparticulation also offers a great control over the release of bioactives from the carrier matrix. The rate and amount of drug release can tailored accordingly with suitable modification to particle type and size, drug loading, kind of drug and microenvironment [125,126]. However, the most important of all factors is the type of polymer used for nanoparticle preparation. The hydrophilic/lipophilic characteristics of a polymer may be manipulated to change the drug release profile as demanded [127].

Application of polymers of biological origin ensures higher biocompatibility, degradability and lower immunogenicity than their synthetic counterparts. Poor physical and mechanical properties associated with biopolymers can be improved through proper use of cross-linkers such as polyethylene glycols or by conjugation with other biomaterials [128]. PLGA or poly(lactide-*co*-glycolic acid) is a US Food and Drug Administration (FDA) approved biopolymer and has garnered unprecedented attention for its biocompatibility and degradability [129]. Modified PLGA nanoparticles prepared through emulsion evaporation technique improved the oral bioavailability of curcumin by several folds in rats [130]. Later studies on PLGA nanoparticles revealed that the mechanism of increasing the bioavailability of payload was through inhibition of P-gp-mediated efflux [131]. PLGA nanoparticles have also been applied as carriers for delivery of different plant biomolecules to specific sites. Curcumin loaded in PLGA-lecithin-PEG nanoparticles were functionalized with RNA aptamers and their *in vitro* therapeutic efficacy were successfully established against colorectal adenocarcinoma cell lines [132]. In one study, ascorbic acid grafted PLGA-b-PEG nanoparticles were developed to transport an alkaloid, galantamine, across the blood-brain-barrier [133]. Sometimes, PLGA were used in combination with other biopolymers for optimizing the release rate and mechanical behaviour of the nanoparticles. Nano-composite particles of PLGA, chitosan and polyacrylamide were synthesized and functionalized with a toxin receptor, CRM197, and ApoproteinE. Evaluation of these nanoparticles containing rosmarinic acid demonstrated promising results in β amyloid damage induced cell lines [134]. Nano-carriers based on PLGA have also been explored as a means of limiting the toxicity associated with anti-neoplastics through co-delivery of quercetin [135]. Other biopolymers such as polylactic acid (PLA), polycaprolactone (PCL) have been tested as safe carriers for quercetin and other biomolecules in various cell lines [136–141]. Unique strategies such as encapsulation of payload in a nanoparticle core while covering the structure with a shell, are being experimented using various combination of biopolymers. Alginate and modified chitosan were focused for encapsulating quercetin in core-shell nanoparticles, and the therapeutic potential of the design was tested in anti-diabetic conditions. This design enabled self-sustained release of quercetin following oral administration and a pronounced hypoglycaemic effect was recorded [142]. A similar core-shell nanoparticle design was developed for encapsulation of resveratrol using zein-pectin combination. The designer biopolymer nanoparticles were intended for improving the antioxidant activity of biomolecule [143].

Facile bio-conjugation of metal nanoparticles is another strategy applied for amplifying the antioxidant potentials of biomolecules. Enhanced antioxidant capacity of bioactive conjugated metal nanoparticles have been studied in various *in vitro* and *in vivo* interfaces [144–147]. Plant bioactives such as anthocyanin and hyperforin were tethered to gold nanosurface, and their efficacy in experimental neuroinflammatory models were confirmed [148,149]. Gold nanoparticles conjugated with polyphenols including EGCG [144], quercetin [145], gallic acid [150], and silybin [151] have found local applications for control of microbial burden in cutaneous wounds. One of the foremost advantage of the gold nanoparticles is that they can be tuned according to their therapeutic applicability, through appropriate design of their size, shape and surface functionalization [152]. Other metals such as silver, selenium, and silica are also being currently explored as platforms for bioactive functionalization [147,153,154]. These biomolecules conversely contribute to the eco-friendly synthesis without any need of noxious chemicals, and offer better stability of the metal nanoparticles [155,156]. Pure plant compounds have also been tested to produce monodisperse nanoparticles in controlled conditions [157], which is a prerequisite in biological applications.

Although more than 10,80,000 studies about nanoparticles have appeared in the literature, only a handful of them reached the clinical trial phase or involve therapeutic use. The primary constraints involve inadequate understanding of the physicochemical properties of the nanomaterials in the biological interfaces [158]. The biological conditions influence the interactions between the plasma constituents and nanoparticles, which consequently underlines and fate and biosafety of the nanoparticles [159,160]. Therefore, the clinical success of the nanoparticles is solely dependent upon major interactions with the plasma proteins.

1.5. Selective fibril-inhibitory strategies

One of the therapeutic approaches to abrogate or mitigate tissue degeneration include inhibition of protein fibrillation and disruption of pre-existing amyloid fibrils. Development of an efficient inhibitor of protein fibrillation is a quite challenge to the drug researchers and scientists. The inhibitor should be able to target the protein fibrillation process in following levels:

(i)   Limit the expression of amyloidogenic proteins

(ii)   Preservation of native protein structures.

(iii) Prevention of aggregation of oligomers directly or indirectly using small molecules.

(iv)   Mitigation of post-pathological symptoms.

Several small molecules [161–163], functionalized polymers [164,165], nanoparticles [166], chaperones [167,168], surfactants [169], peptides [170,171] and drugs have been investigated *in silico* and *in vitro* as possible therapeutics against protein aggregation and amyloid formations. They attempt to suppress the process through a number of mechanisms such as stabilization of native protein structures, destabilization of misfolded aggregates or stimulate rate of re-folding (figure 1.1).

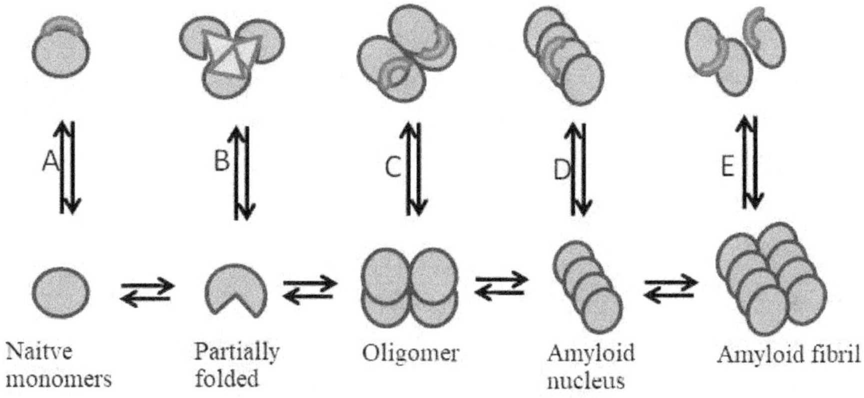

Figure 1.1. Schematic representation of protein fibril inhibition strategies: (A) Native state stabilization (B) Re-folding of polypeptides (C) Off-pathway oligomer formation (D) Disruption of β sheets (E) Disaggregation of amyloid aggregation (Source: International Journal of Biological Macromolecules 2019, 134:1022-1037)

Dyes for instance Congo red, inhibit protein self-assembly formation apart from its biophysical application as a molecular probe in the fibrillation process [172]. Studies proved that Congo red promotes the clearance of huntingtin protein aggregates in both *in vitro* and *in vivo* conditions. Certain molecular tweezers bind to amyloidogenic proteins at specific sites and sequester them into non-toxic oligomers [173]. Application of chaperones is an effective way to enhance the refolding of several proteins. Molecular chaperones such as spectrin assist in re-folding of proteins which are prone to misfolding in physiological stress [174]. Peptides originating from bovine seminal plasma showed chaperone-like activity through inhibition of target proteins such as insulin, lactose dehydrogenase and lactose dehydrogenase under denaturing conditions [175]. Synthetic chaperones based on cholesterol-pullulan nanogels have increased the re-folding efficiency of β amyloid proteins and evaded their aggregation [176]. Another chaperone proSAAS is a product of recombinant technology and has been reported to inhibit fibrillation of α synuclein *in vitro* [177].

Surfactant molecules have revealed to play dual role as inducer and inhibitor of protein fibrillation, at different concentrations [178]. A certain dicationic ester bonded gemini surfactant has been reported to decrease the β sheet content of bovine serum albumin (BSA) fibril through stabilization of the native protein tertiary structure. Micelles of the surfactant further aided in the disruption of the hydrogen bonds occurring within the β sheet strands [179,180]. Typical surfactants including sodium dodecyl sulphate (SDS) and cetyltrimethylammonium bromide (CTAB) inhibit the formation of lysozyme fibrils, but above their critical concentrations [181]. Sulfobetaines, a class of non-detergent surfactants, have also demonstrated effective suppression of both chemical and thermal induced aggregation [182].

Amino acids are the fundamental units of protein chain, and have been explored as inhibitors against the aggregation of diverse amyloidogenic proteins. Of all the amino acids, cysteine and arginine have demonstrated significant fibril inhibitory activity [183,184]. Cysteine diminishes the fibrillation of stem bromelain by stabilizing the native protein structures through hydrogen bond extensions. Another work revealed that cysteine restricted the fibrillation of amyloid β peptide variants through non-covalent thiophilic interactions [185]. The thiophilic interaction appeared between the side chain of the peptide and sulphur motif of the inhibitor, thus inhibiting protein fibrillation. D-amino acids mimic naturally occurring peptides and bind with amyloid β peptide to suppress its fibrillation [186].

With the advancement of biotechnology and molecular biology, diverse range of fibrillation inhibitory peptides have screened and synthesized in the last decade. A number of arginine rich peptides were able to thwart amyloid β fibrillation *in vitro* [187]. Sometimes, short sequences of peptides such as LVFFA and KLVFF were designed to mimic amyloid β structure and effectively inhibit its fibrillation [188]. Peptidomimetic-type fibril inhibitors can also be synthesized by linking hydrophobic dipeptides to a sugar moiety. These inhibitors are efficacious at very low concentrations, and their bio-activities are reported to be target protein-specific [189]. In 2018, a transthyretin mimicking cyclic peptide was developed which exhibited potent therapeutic activity by suppressing the fibrillation of amyloid β and amylin polypeptides [190].

A large number of other molecules such as nicotine [191], hemin [192], rifampicin [193], melatonin [194], aspirin [195] and fullerene [196] have also been reported to repress the fibrillation process. Apart from aspirin, other NSAIDs decrease the physiological production of amyloid β peptide by interacting with γ secretase complex [197]. The well-known anti-Parkinson drug levodopa proved therapeutically effective in systemic amyloidosis [198] and disintegrated mature serum albumin fibrils *in vitro* [199]. Another anti-Parkinson drug selegiline preferably delays the lag phase and blocks the toxic aggregate formation pathways [200]. The effect of a second-line anti-tuberculosis drug, capreomycin, on fibrillation of insulin has been explored quite recently. In contrast to selegiline, capreomycin dose-dependently suppresses the fibril elongating step rather than the lag phase of protein fibrillation [201].

While revolutionizing the arenas of medicine, nanoparticles have been recognized as promising tools in protein interface. It is now established that protein molecules are rapidly adsorbed on the nanoparticle surface and experience perturbation in their structural stability [202]. Various properties (size, surface charge, chemical makeup) of the nanoparticles influence the protein self-assembly, and alter important biological processes [203]. While nanoparticles of cerium oxide, titanium dioxide and carbon nanotubes may aggravate protein misfolding related disease conditions, other nanoparticles such as maghemite, N-acetyl-L-cysteine capped quantum dots, fullerene, conjugated gold nanoparticles have exhibited significantly high amyloidogenic capacity. Certain nanoparticles of zinc oxide and silica have been reported to be lack any major effect on amyloid β fibrillation [204,205]. Interestingly, a dual effect on amyloid β fibril formation was observed in case of amine-modified polystyrene nanoparticles. Low particle surface area caused an acceleration of the fibrillation process through shortening of the lag phase, and the opposite effect prevailed in case of high particle surface area [206]. Though there are extensive reports on the effects of nanomaterials on protein fibrillation kinetics, there are many factors that were ignored and must be considered for clinical applicability as nanomedicines. One of the most critical factor is the effect of instantaneous binding of non-amyloidogenic proteins upon the nanosurface, thus forming a "corona" [207]. Therefore, what the amyloidogenic proteins perceive *in vivo* is the nanoparticle-protein corona form, rather than the bare or pristine nanoparticles. This has led to reduced access of amyloid β peptides to the nanosurface, consequently may decrease the effectiveness of the nanomedicines.

Most of the materials investigated till now have been met with limited successes when tested *in vivo,* and few have passed the clinical trials [208]. Cyclohexane-1,2,3,4,5,6-hexaol and methylthioninium chloride have reached to clinical trials for Alzheimer's Disease, though the studies were discontinued owing to lack of promising data [209,210]. This scenario therefore demands a careful and comprehensive design of fibril inhibitor candidates that apprehends the failures and missing links in the earlier works.

1.6. Trends in fibril-inhibitory studies

In the search for fibrillation modulators, natural products have exhibited promising results. Plant bioactives have a clear advantage over others, as they are frequently consumed as part of daily diet. These bioactives are also utilized as rich sources for various structural backbones in rational designing of multi-functional anti-amyloid drugs [211]. About 72 such compounds, comprising of polyphenols, flavonoids, anthraquinones, terpenes and alkaloids have been identified to impede or modulate protein aggregation [60]. Natural amyloid inhibitors present in a healthy diet include oleuropein and oleocanthal available from olive oil, curcumin from turmeric, resveratrol from red wine [212], myricetin from green tea, naringin and kaempferol from different fruits [213], and ellagic acid obtained from berries [214]. Other polyphenolics such as caffeic acid [215] and rosmarinic acid [216] obtained from certain herbs, and genistein found in leguminous plants [217] have shown significant fibril inhibitory efficacy against amyloidogenic proteins.

The specific steps of protein fibrillation and amyloid species that are targeted by the plant bioactives, as well as the biomolecular processes related to amyloid induced tissue toxicity are not yet completely understood. Recent studies however are focused on elucidating the intricate chemical mechanisms of fibril inhibition at a molecular level. Different works have revealed that a particular bioactive can suppress fibrillation through formation of covalent bonds [218–220] and/or non-covalent interactions [221–223] with the amino acid residues of the amyloidogenic proteins, thereby modulating one or all the steps of fibrillation.

Several *in silico* studies have revealed that curcumin binds to the hydrophobic cores of all amyloid assemblies irrespective of the type and structure of the target protein [224– 226]. The curcumin molecule targets the cross β sheet spine of all amyloid fibrils as well as α helical oligomers which assemble during the early stages of protein fibrillation [227]. These non-covalent interactions hence play a major role in asserting the fibril delaying and inhibitory effects of curcumin. Another polyphenol EGCG extends hydrophobic and hydrogen bonds to prevent the formation of fibrillar structures and also remodel the matured fibrils into less toxic aggregates [222,228,229]. For suppression of amylin misfolding, non-specific (hydrogen/ hydrophobic) interactions mediate the bioactivity of EGCG. However, in case of PAP (prostatic acid phosphatase-cleaved amyloid precursor peptide) fibrillation inhibition, specific interactions play the lead role [230]. The specificity and types of non-covalent interactions extended by the bioactive are therefore influenced by the target protein or the stage of aggregation [231–234].

Certain flavonoids and catechol containing phenolics interact with amyloid proteins through o-quinone mediated covalent adduct formation. A series of NMR analyses and mass spectroscopies suggested that gallic acid ester of EGCG mediates adduct formation with the lysine residues of PAP [230]. Taxifolin triggers cross β sheet instability and repress β amyloid fibrillation under aerobic conditions. Site specific mutagenesis and mass spectroscopy confirmed that o-quinone-lysine covalent adduct formation through Michael addition was responsible for anti-amyloid capacity of taxifolin [218]. Catechol-type compounds such as myricetin and catechol also exhibited anti-amyloid action on β amyloid peptides through similar covalent interactions [218]. Other fibrillation-discouraging covalent interactions have been reported between nucleophilic groups of amyloidogenic proteins and electrophilic scaffolds (such as aldehydes) present in cinnamaldehyde, epicatechin [235] and oleocanthal [236].

Some plant bioactives demonstrated their anti-amyloid effects through alteration of fibril formation pathway, favouring the generation of inhibitor-bound soluble and insoluble protein aggregates [237–240]. Such aggregates display certain biochemical and biophysical properties such as wide range of size distributions, higher stability in denaturing environment and inertness to cell viability [233,240]. These properties separate them from the typical pre-fibrillar oligomers which are the main contributors of amyloid induced toxicity [241–243]. Baicalein has been reported to produce off-pathway species of stable α synuclein oligomers through covalent adduct formation. These oligomers further inhibited the monomeric protein fibril formation in a concentration dependant manner [244,245]. Bioactives such as resveratrol, nordihydroguaiaretic acid and myricetin guide preformed β amyloid fibrils towards stable, off-pathway protein aggregates having lower cytotoxicity in comparison to mature fibrils [240]. However, in contrast of baicalein, these bioactives form inhibitor-aggregate complexes through non-covalent interactions. Polyphenol derivatives of exifone and rafigallol suppress tau protein fibril assembly by retaining the maximum fraction of tau protein in stable, soluble form compared to untreated tau protein [246]. Betulinic acid, a triterpenoid molecule, have shown to accelerate the formation β amyloid fibrils, thereby inhibiting the formation of toxic, soluble, low molecular weight oligomers. Betulinic acid induced acceleration of fibril formation is an exclusive example of "on-pathway" process [247]. Another terpenoid, andrographolide, is a well-known alleviator of several human pathologies including Alzheimer's disease [248,249]. However, the effects of this biomolecule in protein fibril interface is not investigated.

Plant bioactives are not without limitations. Although the initial research on the compounds appeared promising, the fibril inhibitory functions of most of these biomolecules have not been validated *in vivo*. A considerable number of plant-based amyloid inhibitors, such as quercetin, are highly lipophilic, and therefore not ideal for drug delivery. Applications of such molecules are also limited due to their low stability in physiological environments [250,251]. Moreover, administration of large doses of quercetin cannot be implemented to address its bioavailability issues, as the biomolecule demonstrates significant cytotoxic properties.

Newer synthetic compounds have come up relatively recently for specific protein fibrillation inhibition. Picolinic acid derivative M1 appeared as one promising compound for insulin fibrillation inhibition [252]. Caffeic acid congener rosmarinic acid from basil (*Ocimum basilicum*) was proposed as one albumin misfolding and aggregation inhibitor [253]. Well known polyphenol morin, from guava (*Psidium guajava*) leaves was reported as useful against insulin fibrillation [254]. *Rosa damascene*, gallates and kaempferol glycosides were highly potent against α-synuclein fibrillation [255]. Currently, there is an intensive search for natural compounds which could be useful for inhibition of fibrillation and oligomerization of disease related proteins. Studies on natural compounds in protein fibrillation interface are so far fascinating, and incite further exploration into plant bioactives.

1.7. References

[1]   K. Mazmanian, K. Sargsyan, C. Lim, How the Local Environment of Functional Sites Regulates Protein Function, J. Am. Chem. Soc. 142 (2020) 9861–9871. doi:10.1021/jacs.0c02430.

[2]   P.C. Ke, M.-A. Sani, F. Ding, A. Kakinen, I. Javed, F. Separovic, T.P. Davis, R. Mezzenga, Implications of peptide assemblies in amyloid diseases, Chem. Soc. Rev. 46 (2017) 6492–6531. doi:10.1039/C7CS00372B.

[3]   S.K. Chaturvedi, M.K. Siddiqi, P. Alam, R.H. Khan, Protein misfolding and aggregation: Mechanism, factors and detection, Process Biochem. 51 (2016) 1183–1192. doi:10.1016/j.procbio.2016.05.015.

[4]   H. Ramshini, M. mohammad-zadeh, A. Ebrahim-Habibi, Inhibition of amyloid fibril formation and cytotoxicity by a chemical analog of Curcumin as a stable inhibitor, Int. J. Biol. Macromol. 78 (2015) 396–404. doi:10.1016/j.ijbiomac.2015.04.038.

[5]   K.-N. Liu, C. Lai, Y. Lee, S.-N. Wang, R.P.-Y. Chen, J. Jan, H. Liu, S.S.-S. Wang, Curcumin's pre-incubation temperature affects its inhibitory potency toward amyloid fibrillation and fibril-induced cytotoxicity of lysozyme, Biochim. Biophys. Acta - Gen. Subj. 1820 (2012) 1774–1786. doi:10.1016/j.bbagen.2012.07.012.

[6]   J. He, Y. Wang, A.K. Chang, L. Xu, N. Wang, X. Chong, H. Li, B. Zhang, G.W. Jones, Y. Song, Myricetin prevents fibrillogenesis of hen egg white lysozyme, J. Agric. Food Chem. 62 (2014) 9442–9449. doi:10.1021/jf5025449.

[7]   L. Holm, C. Sander, Mapping the Protein Universe, Science (80-.). 273 (1996) 595 LP – 602. doi:10.1126/science.273.5275.595.

[8]   C.M. Dobson, Protein folding and misfolding, Nature. 426 (2003) 884–890. doi:10.1038/nature02261.

[9]   B. Alberts, A. Johnson, J. Lewis, M. Raff, K. Roberts, P. Walter, The shape and structure of proteins, in: Mol. Biol. Cell. 4th Ed., Garland Science, 2002.

[10]   J. Tan, J. Zhang, C. Li, Y. Luo, S. Ye, Ultrafast energy relaxation dynamics of amide I vibrations coupled with protein-bound water molecules, Nat. Commun. 10 (2019) 1010. doi:10.1038/s41467-019-08899-3.

[11]   M.S. Cheung, A.E. García, J.N. Onuchic, Protein folding mediated by solvation: Water expulsion and formation of the hydrophobic core occur after the structural collapse, Proc. Natl. Acad. Sci. 99 (2002) 685 LP – 690. doi:10.1073/pnas.022387699.

[12]   J.N. Onuchic, P.G. Wolynes, Theory of protein folding, Curr. Opin. Struct. Biol. 14 (2004) 70–75. doi:10.1016/j.sbi.2004.01.009.

[13]   N.H. Joh, A. Min, S. Faham, J.P. Whitelegge, D. Yang, V.L. Woods, J.U. Bowie, Modest stabilization by most hydrogen-bonded side-chain interactions in membrane proteins, Nature. 453 (2008) 1266–1270. doi:10.1038/nature06977.

[14]   S.E. Radford, What's new in protein folding? EMBO Workshop: Protein folding and misfolding inside and outside the cell, Fold. Des. 3 (1998) R59–R63. doi:10.1016/S1359-0278(98)00022-4.

[15]   J. Gandhi, A.C. Antonelli, A. Afridi, S. Vatsia, G. Joshi, V. Romanov, I.V.J. Murray, S.A. Khan, Protein misfolding and aggregation in neurodegenerative diseases: a review of pathogeneses, novel detection strategies, and potential therapeutics, Rev. Neurosci. 30 (2019) 339–358. doi:10.1515/revneuro-2016-0035.

[16]   J. Tyedmers, A. Mogk, B. Bukau, Cellular strategies for controlling protein aggregation, Nat. Rev. Mol. Cell Biol. 11 (2010) 777–788. doi:10.1038/nrm2993.

[17]   Q. Wan, D. Song, H. Li, M. He, Stress proteins: the biological functions in virus infection, present and challenges for target-based antiviral drug development, Signal Transduct. Target. Ther. 5 (2020) 125. doi:10.1038/s41392-020-00233-4.

[18]   L.P. Coyne, X.J. Chen, Consequences of inner mitochondrial membrane protein misfolding, Mitochondrion. 49 (2019) 46–55. doi:10.1016/j.mito.2019.06.001.

[19]   N.B. Pham, W.S. Meng, Protein aggregation and immunogenicity of biotherapeutics, Int. J. Pharm. 585 (2020) 119523. doi:10.1016/j.ijpharm.2020.119523.

[20]   M. Zaman, A.N. Khan, Wahiduzzaman, S.M. Zakariya, R.H. Khan, Protein misfolding, aggregation and mechanism of amyloid cytotoxicity: An overview and therapeutic strategies to inhibit aggregation, Int. J. Biol. Macromol. 134 (2019) 1022–1037. doi:10.1016/j.ijbiomac.2019.05.109.

[21]   Y. Misumi, Y. Ando, M. Ueda, K. Obayashi, H. Jono, Y. Su, T. Yamashita, M. Uchino, Chain reaction of amyloid fibril formation with induction of basement

membrane in familial amyloidotic polyneuropathy, J. Pathol. 219 (2009) 481–490. doi:10.1002/path.2618.

[22]   A.T. Petkova, Y. Ishii, J.J. Balbach, O.N. Antzutkin, R.D. Leapman, F. Delaglio, R. Tycko, A structural model for Alzheimer's β-amyloid fibrils based on experimental constraints from solid state NMR, Proc. Natl. Acad. Sci. 99 (2002) 16742–16747. doi:10.1073/pnas.262663499.

[23]   S.J. Roeters, A. Iyer, G. Pletikapić, V. Kogan, V. Subramaniam, S. Woutersen, Evidence for Intramolecular Antiparallel Beta-Sheet Structure in Alpha-Synuclein Fibrils from a Combination of Two-Dimensional Infrared Spectroscopy and Atomic Force Microscopy, Sci. Rep. 7 (2017) 41051. doi:10.1038/srep41051.

[24]   E. Barbu, M. Joly, The globular-fibrous protein transformation, Discuss. Faraday Soc. 13 (1953) 77–93. doi:10.1039/DF9531300077.

[25]   D.N. Brems, L.A. Alter, M.J. Beckage, R.E. Chance, R.D. DiMarchi, L.K. Green, H.B. Long, A.H. Pekar, J.E. Shields, B.H. Frank, Altering the association properties of insulin by amino acid replacement, Protein Eng. Des. Sel. 5 (1992) 527–533. doi:10.1093/protein/5.6.527.

[26]   J.R. Alford, B.S. Kendrick, J.F. Carpenter, T.W. Randolph, High Concentration Formulations of Recombinant Human Interleukin-1 Receptor Antagonist: II. Aggregation Kinetics, J. Pharm. Sci. 97 (2008) 3005–3021. doi:10.1002/jps.21205.

[27]   R. Raffen, L.J. Dieckman, M. Szpunar, C. Wunschl, P.R. Pokkuluri, P. Dave, P. Wilkins Stevens, X. Cai, M. Schiffer, F.J. Stevens, Physicochemical consequences of amino acid variations that contribute to fibril formation by immunoglobulin light chains, Protein Sci. 8 (1999) 509–517. doi:10.1110/ps.8.3.509.

[28]   D. Canet, A.M. Last, P. Tito, M. Sunde, A. Spencer, D.B. Archer, C. Redfield, C. V Robinson, C.M. Dobson, Local cooperativity in the unfolding of an amyloidogenic variant of human lysozyme, Nat. Struct. Biol. 9 (2002) 308–315. doi:10.1038/nsb768.

[29]   B.S. Kendrick, J.F. Carpenter, J.L. Cleland, T.W. Randolph, A transient expansion of the native state precedes aggregation of recombinant human interferon-γ, Proc. Natl. Acad. Sci. 95 (1998) 14142 LP – 14146. doi:10.1073/pnas.95.24.14142.

[30]   P. Hammarström, X. Jiang, A.R. Hurshman, E.T. Powers, J.W. Kelly, Sequence-

dependent denaturation energetics: A major determinant in amyloid disease diversity, Proc. Natl. Acad. Sci. 99 (2002) 16427 LP – 16432. doi:10.1073/pnas.202495199.

[31] N. CHAYEN, Methods for separating nucleation and growth in protein crystallisation, Prog. Biophys. Mol. Biol. 88 (2005) 329–337. doi:10.1016/j.pbiomolbio.2004.07.007.

[32] J. Philo, T. Arakawa, Mechanisms of Protein Aggregation, Curr. Pharm. Biotechnol. 10 (2009) 348–351. doi:10.2174/138920109788488932.

[33] M.I. Smith, J.S. Sharp, C.J. Roberts, Nucleation and Growth of Insulin Fibrils in Bulk Solution and at Hydrophobic Polystyrene Surfaces, Biophys. J. 93 (2007) 2143–2151. doi:10.1529/biophysj.107.105338.

[34] L. Nault, P. Guo, B. Jain, Y. Bréchet, F. Bruckert, M. Weidenhaupt, Human insulin adsorption kinetics, conformational changes and amyloidal aggregate formation on hydrophobic surfaces, Acta Biomater. 9 (2013) 5070–5079. doi:10.1016/j.actbio.2012.09.025.

[35] H.J. Dyson, P.E. Wright, Intrinsically unstructured proteins and their functions, Nat. Rev. Mol. Cell Biol. 6 (2005) 197–208. doi:10.1038/nrm1589.

[36] B.B. Holmes, S.L. DeVos, N. Kfoury, M. Li, R. Jacks, K. Yanamandra, M.O. Ouidja, F.M. Brodsky, J. Marasa, D.P. Bagchi, P.T. Kotzbauer, T.M. Miller, D. Papy-Garcia, M.I. Diamond, Heparan sulfate proteoglycans mediate internalization and propagation of specific proteopathic seeds, Proc. Natl. Acad. Sci. 110 (2013) E3138–E3147. doi:10.1073/pnas.1301440110.

[37] M.G. Iadanza, M.P. Jackson, E.W. Hewitt, N.A. Ranson, S.E. Radford, A new era for understanding amyloid structures and disease, Nat. Rev. Mol. Cell Biol. 19 (2018) 755–773. doi:10.1038/s41580-018-0060-8.

[38] N.K. Bart, L. Thomas, D. Korczyk, J.J. Atherton, G.J. Stewart, D. Fatkin, Amyloid Cardiomyopathy, Hear. Lung Circ. 29 (2020) 575–583. doi:10.1016/j.hlc.2019.11.019.

[39] M.F. Knauer, B. Soreghan, D. Burdick, J. Kosmoski, C.G. Glabe, Intracellular accumulation and resistance to degradation of the Alzheimer amyloid A4/β protein, Proc. Natl. Acad. Sci. U. S. A. 89 (1992) 7437–7441. doi:10.1073/pnas.89.16.7437.

[40] C. Haass, C.A. Lemere, A. Capell, M. Citron, P. Seubert, D. Schenk, L. Lannfelt, D.J. Selkoe, The Swedish mutation causes early-onset Alzheimer's disease by β-secretase

cleavage within the secretory pathway, Nat. Med. 1 (1995) 1291–1296. doi:10.1038/nm1295-1291.

[41]   F.. De Beer, E. Fagan, G.R.. Hughes, R.. Mallya, J.. Lanham, M.. Pepys, SERUM AMYLOID-A PROTEIN CONCENTRATION IN INFLAMMATORY DISEASES AND ITS RELATIONSHIP TO THE INCIDENCE OF REACTIVE SYSTEMIC AMYLOIDOSIS, Lancet. 320 (1982) 231–234. doi:10.1016/S0140-6736(82)90321-X.

[42]   M.-H. Yu, X. Li, Q. Li, S.-J. Mo, Y. Ni, F. Han, Y.-B. Wang, Y.-X. Tu, SAA1 increases NOX4/ROS production to promote LPS-induced inflammation in vascular smooth muscle cells through activating p38MAPK/NF-κB pathway, BMC Mol. Cell Biol. 20 (2019) 15. doi:10.1186/s12860-019-0197-0.

[43]   M.S. Wolfe, Tau Mutations in Neurodegenerative Diseases, J. Biol. Chem. 284 (2009) 6021–6025. doi:10.1074/jbc.R800013200.

[44]   M.H. Polymeropoulos, Mutation in the α-Synuclein Gene Identified in Families with Parkinson's Disease, Science (80-.). 276 (1997) 2045–2047. doi:10.1126/science.276.5321.2045.

[45]   E.M. Rocha, B. De Miranda, L.H. Sanders, Alpha-synuclein: Pathology, mitochondrial dysfunction and neuroinflammation in Parkinson's disease, Neurobiol. Dis. 109 (2018) 249–257. doi:10.1016/j.nbd.2017.04.004.

[46]   G.J. Miroy, Z. Lai, H.A. Lashuel, S.A. Peterson, C. Strang, J.W. Kelly, Inhibiting transthyretin amyloid fibril formation via protein stabilization, Proc. Natl. Acad. Sci. 93 (1996) 15051–15056. doi:10.1073/pnas.93.26.15051.

[47]   Y. Furukawa, N. Nukina, Functional diversity of protein fibrillar aggregates from physiology to RNA granules to neurodegenerative diseases, Biochim. Biophys. Acta - Mol. Basis Dis. 1832 (2013) 1271–1278. doi:10.1016/j.bbadis.2013.04.011.

[48]   A.C. Gill, A.R. Castle, Chapter 2 - The cellular and pathologic prion protein, in: M. Pocchiari, J.B.T.-H. of C.N. Manson (Eds.), Hum. Prion Dis., Elsevier, 2018: pp. 21–44. doi:https://doi.org/10.1016/B978-0-444-63945-5.00002-7.

[49]   E.H. Koo, P.T. Lansbury, J.W. Kelly, Amyloid diseases: Abnormal protein aggregation in neurodegeneration, Proc. Natl. Acad. Sci. 96 (1999) 9989–9990. doi:10.1073/pnas.96.18.9989.

[50] J. Juárez, S.G. López, A. Cambón, P. Taboada, V. Mosquera, Influence of electrostatic interactions on the fibrillation process of human serum albumin., J. Phys. Chem. B. 113 (2009) 10521–10529. doi:10.1021/jp902224d.

[51] S. Bag, R. Mitra, S. DasGupta, S. Dasgupta, Inhibition of Human Serum Albumin Fibrillation by Two-Dimensional Nanoparticles, J. Phys. Chem. B. 121 (2017) 5474–5482. doi:10.1021/acs.jpcb.7b01289.

[52] L. Xu, Y.X. Hu, Y.C. Li, L. Zhang, H.X. Ai, H.S. Liu, Y.F. Liu, Y.L. Sang, Study on the interaction of tussilagone with human serum albumin (HSA) by spectroscopic and molecular docking techniques, J. Mol. Struct. 1149 (2017) 645–654. doi:10.1016/j.molstruc.2017.08.039.

[53] P. Taboada, S. Barbosa, E. Castro, V. Mosquera, Amyloid fibril formation and other aggregate species formed by human serum albumin association., J. Phys. Chem. B. 110 (2006) 20733–20736. doi:10.1021/jp064861r.

[54] R.S. Solecki, Shanidar IV, a Neanderthal Flower Burial in Northern Iraq, Science (80-.). 190 (1975) 880–881. doi:10.1126/science.190.4217.880.

[55] V.P. Kamboj, Herbal medicine, Curr. Sci. 78 (2000) 35–39. http://www.jstor.org/stable/24103844.

[56] V.K. Srivastav, C. Egbuna, M. Tiwari, Plant secondary metabolites as lead compounds for the production of potent drugs, in: C. Egbuna, S. Kumar, J.C. Ifemeje, S.M. Ezzat, S.B.T.-P. as L.C. for N.D.D. Kaliyaperumal (Eds.), Phytochem. as Lead Compd. New Drug Discov., Elsevier, 2020: pp. 3–14. doi:10.1016/B978-0-12-817890-4.00001-9.

[57] J. Andersen-Ranberg, K.T. Kongstad, M.T. Nielsen, N.B. Jensen, I. Pateraki, S.S. Bach, B. Hamberger, P. Zerbe, D. Staerk, J. Bohlmann, B.L. Møller, B. Hamberger, Expanding the Landscape of Diterpene Structural Diversity through Stereochemically Controlled Combinatorial Biosynthesis, Angew. Chemie Int. Ed. 55 (2016) 2142–2146. doi:10.1002/anie.201510650.

[58] S.H. Omar, C.J. Scott, A.S. Hamlin, H.K. Obied, The protective role of plant biophenols in mechanisms of Alzheimer's disease, J. Nutr. Biochem. 47 (2017) 1–20. doi:10.1016/j.jnutbio.2017.02.016.

[59] L. Cassidy, F. Fernandez, J.B. Johnson, M. Naiker, A.G. Owoola, D.A. Broszczak,

Oxidative stress in alzheimer's disease: A review on emergent natural polyphenolic therapeutics, Complement. Ther. Med. 49 (2020) 102294. doi:10.1016/j.ctim.2019.102294.

[60] P. Velander, L. Wu, F. Henderson, S. Zhang, D.R. Bevan, B. Xu, Natural product-based amyloid inhibitors, Biochem. Pharmacol. 139 (2017) 40–55. doi:10.1016/j.bcp.2017.04.004.

[61] A.P. Singh, R. Singh, S.S. Verma, V. Rai, C.H. Kaschula, P. Maiti, S.C. Gupta, Health benefits of resveratrol: Evidence from clinical studies, Med. Res. Rev. 39 (2019) 1851–1891. doi:10.1002/med.21565.

[62] E.-J. Seo, N. Fischer, T. Efferth, Phytochemicals as inhibitors of NF-κB for treatment of Alzheimer's disease, Pharmacol. Res. 129 (2018) 262–273. doi:10.1016/j.phrs.2017.11.030.

[63] M.-J. Kim, A.-R. Seong, J.-Y. Yoo, C.-H. Jin, Y.-H. Lee, Y.J. Kim, J. Lee, W.J. Jun, H.-G. Yoon, Gallic acid, a histone acetyltransferase inhibitor, suppresses β-amyloid neurotoxicity by inhibiting microglial-mediated neuroinflammation, Mol. Nutr. Food Res. 55 (2011) 1798–1808. doi:10.1002/mnfr.201100262.

[64] A. Rajeswari, M. Sabesan, Inhibition of monoamine oxidase-B by the polyphenolic compound, curcumin and its metabolite tetrahydrocurcumin, in a model of Parkinson's disease induced by MPTP neurodegeneration in mice, Inflammopharmacology. 16 (2008) 96–99. doi:10.1007/s10787-007-1614-0.

[65] O. Weinreb, S. Mandel, T. Amit, M.B.H. Youdim, Neurological mechanisms of green tea polyphenols in Alzheimer's and Parkinson's diseases, J. Nutr. Biochem. 15 (2004) 506–516. doi:10.1016/j.jnutbio.2004.05.002.

[66] Z.-H. Wang, J.-L. Zhang, Y.-L. Duan, Q.-S. Zhang, G.-F. Li, D.-L. Zheng, MicroRNA-214 participates in the neuroprotective effect of Resveratrol via inhibiting α-synuclein expression in MPTP-induced Parkinson's disease mouse, Biomed. Pharmacother. 74 (2015) 252–256. doi:10.1016/j.biopha.2015.08.025.

[67] S. Sen Singh, S.N. Rai, H. Birla, W. Zahra, A.S. Rathore, S.P. Singh, NF-κB-Mediated Neuroinflammation in Parkinson's Disease and Potential Therapeutic Effect of Polyphenols, Neurotox. Res. 37 (2020) 491–507. doi:10.1007/s12640-019-00147-2.

[68] L. Bonilla-Ramirez, M. Jimenez-Del-Rio, C. Velez-Pardo, Low doses of paraquat and polyphenols prolong life span and locomotor activity in knock-down parkin Drosophila melanogaster exposed to oxidative stress stimuli: Implication in autosomal recessive juvenile Parkinsonism, Gene. 512 (2013) 355–363. doi:10.1016/j.gene.2012.09.120.

[69] B. Pang, Treatment of refractory diabetic gastroparesis: Western medicine and traditional Chinese medicine therapies, World J. Gastroenterol. 20 (2014) 6504. doi:10.3748/wjg.v20.i21.6504.

[70] Y. Fu, J. Luo, Z. Jia, W. Zhen, K. Zhou, E. Gilbert, D. Liu, Baicalein Protects against Type 2 Diabetes via Promoting Islet β-Cell Function in Obese Diabetic Mice, Int. J. Endocrinol. 2014 (2014) 846742. doi:10.1155/2014/846742.

[71] L. Yang, H. Sun, L. Wu, X. Guo, H. Dou, M.O.M. Tso, L. Zhao, S. Li, Baicalein Reduces Inflammatory Process in a Rodent Model of Diabetic Retinopathy, Invest. Ophthalmol. Vis. Sci. 50 (2009) 2319–2327. doi:10.1167/iovs.08-2642.

[72] W. Wang, C. Wang, X.Q. Ding, Y. Pan, T.T. Gu, M.X. Wang, Y.L. Liu, F.M. Wang, S.J. Wang, L.D. Kong, Quercetin and allopurinol reduce liver thioredoxin-interacting protein to alleviate inflammation and lipid accumulation in diabetic rats, Br. J. Pharmacol. 169 (2013) 1352–1371. doi:10.1111/bph.12226.

[73] A.D. Kandhare, K.S. Raygude, V. Shiva Kumar, A.R. Rajmane, A. Visnagri, A.E. Ghule, P. Ghosh, S.L. Badole, S.L. Bodhankar, Ameliorative effects quercetin against impaired motor nerve function, inflammatory mediators and apoptosis in neonatal streptozotocin-induced diabetic neuropathy in rats, Biomed. Aging Pathol. 2 (2012) 173–186. doi:10.1016/j.biomag.2012.10.002.

[74] S.-J. Kang, J.-H.Y. Park, H.-N. Choi, J.-I. Kim, α-glucosidase inhibitory activities of myricetin in animal models of diabetes mellitus, Food Sci. Biotechnol. 24 (2015) 1897–1900. doi:10.1007/s10068-015-0249-y.

[75] X.-K. Fang, J. Gao, D.-N. Zhu, Kaempferol and quercetin isolated from Euonymus alatus improve glucose uptake of 3T3-L1 cells without adipogenesis activity, Life Sci. 82 (2008) 615–622. doi:10.1016/j.lfs.2007.12.021.

[76] L. Xu, Y. Li, Y. Dai, J. Peng, Natural products for the treatment of type 2 diabetes

mellitus: Pharmacology and mechanisms, Pharmacol. Res. 130 (2018) 451–465. doi:10.1016/j.phrs.2018.01.015.

[77] L. Guan, H. Feng, D. Gong, X. Zhao, L. Cai, Q. Wu, B. Yuan, M. Yang, J. Zhao, Y. Zou, Genipin ameliorates age-related insulin resistance through inhibiting hepatic oxidative stress and mitochondrial dysfunction, Exp. Gerontol. 48 (2013) 1387–1394. doi:10.1016/j.exger.2013.09.001.

[78] X. Wang, R. Liu, W. Zhang, X. Zhang, N. Liao, Z. Wang, W. Li, X. Qin, C. Hai, Oleanolic acid improves hepatic insulin resistance via antioxidant, hypolipidemic and anti-inflammatory effects, Mol. Cell. Endocrinol. 376 (2013) 70–80. doi:10.1016/j.mce.2013.06.014.

[79] X. Ji, C. Li, Y. Ou, N. Li, K. Yuan, G. Yang, X. Chen, Z. Yang, B. Liu, W.W. Cheung, L. Wang, R. Huang, T. Lan, Andrographolide ameliorates diabetic nephropathy by attenuating hyperglycemia-mediated renal oxidative stress and inflammation via Akt/NF-κB pathway, Mol. Cell. Endocrinol. 437 (2016) 268–279. doi:10.1016/j.mce.2016.06.029.

[80] Y.S. Lee, W.S. Kim, K.H. Kim, M.J. Yoon, H.J. Cho, Y. Shen, J.-M. Ye, C.H. Lee, W.K. Oh, C.T. Kim, C. Hohnen-Behrens, A. Gosby, E.W. Kraegen, D.E. James, J.B. Kim, Berberine, a Natural Plant Product, Activates AMP-Activated Protein Kinase With Beneficial Metabolic Effects in Diabetic and Insulin-Resistant States, Diabetes. 55 (2006) 2256–2264. doi:10.2337/db06-0006.

[81] J.-H. Kang, G. Tsuyoshi, I.-S. Han, T. Kawada, Y.M. Kim, R. Yu, Dietary Capsaicin Reduces Obesity-induced Insulin Resistance and Hepatic Steatosis in Obese Mice Fed a High-fat Diet, Obesity. 18 (2010) 780–787. doi:10.1038/oby.2009.301.

[82] S. Choi, Y. Choi, Y. Choi, S. Kim, J. Jang, T. Park, Piperine reverses high fat diet-induced hepatic steatosis and insulin resistance in mice, Food Chem. 141 (2013) 3627–3635. doi:10.1016/j.foodchem.2013.06.028.

[83] A. Papadimitriou, K.C. Silva, E.B.M.I. Peixoto, C.M. Borges, J.M. Lopes de Faria, J.B. Lopes de Faria, Theobromine increases NAD + /Sirt-1 activity and protects the kidney under diabetic conditions, Am. J. Physiol. Physiol. 308 (2015) F209–F225. doi:10.1152/ajprenal.00252.2014.

[84]   A.P. Demidowich, A.I. Davis, N. Dedhia, J.A. Yanovski, Colchicine to decrease
       NLRP3-activated inflammation and improve obesity-related metabolic dysregulation,
       Med. Hypotheses. 92 (2016) 67–73. doi:10.1016/j.mehy.2016.04.039.

[85]   M. Cong, W. Zhao, T. Liu, P. Wang, X. Fan, Q. Zhai, X. Bao, D. Zhang, H. You, T.
       Kisseleva, D.A. Brenner, J. Jia, H. Zhuang, Protective effect of human serum amyloid
       P on CCl4-induced acute liver injury in mice, Int. J. Mol. Med. 40 (2017) 454–464.
       doi:10.3892/ijmm.2017.3028.

[86]   P. Zhang, C. Hölscher, X. Ma, Therapeutic potential of flavonoids in spinal cord
       injury, Rev. Neurosci. 28 (2017) 87. doi:10.1515/revneuro-2016-0053.

[87]   G. Sebastiani, Chronic hepatitis C and liver fibrosis, World J. Gastroenterol. 20 (2014)
       11033. doi:10.3748/wjg.v20.i32.11033.

[88]   I. of Medicine, Hepatitis and Liver Cancer, National Academies Press, Washington,
       D.C., 2010. doi:10.17226/12793.

[89]   B. Tang, H. Qiao, F. Meng, X. Sun, Glycyrrhizin attenuates endotoxin- induced acute
       liver injury after partial hepatectomy in rats, Brazilian J. Med. Biol. Res. 40 (2007)
       1637–1646. doi:10.1590/S0100-879X2006005000173.

[90]   S.J. Polyak, C. Morishima, V. Lohmann, S. Pal, D.Y.W. Lee, Y. Liu, T.N. Graf, N.H.
       Oberlies, Identification of hepatoprotective flavonolignans from silymarin, Proc. Natl.
       Acad. Sci. 107 (2010) 5995–5999. doi:10.1073/pnas.0914009107.

[91]   M. Muthumani, S.M. Prabu, Silibinin potentially protects arsenic-induced oxidative
       hepatic dysfunction in rats, Toxicol. Mech. Methods. 22 (2012) 277–288.
       doi:10.3109/15376516.2011.647113.

[92]   J. Feher, G. Lengyel, Silymarin in the Prevention and Treatment of Liver Diseases and
       Primary Liver Cancer, Curr. Pharm. Biotechnol. 13 (2012) 210–217.
       doi:10.2174/138920112798868818.

[93]   M. Kebieche, Z. Lakroun, M. Lahouel, J. Bouayed, Z. Meraihi, R. Soulimani,
       Evaluation of epirubicin-induced acute oxidative stress toxicity in rat liver cells and
       mitochondria, and the prevention of toxicity through quercetin administration, Exp.
       Toxicol. Pathol. 61 (2009) 161–167. doi:10.1016/j.etp.2008.06.002.

[94]   D. Ghosh, S. Ghosh, S. Sarkar, A. Ghosh, N. Das, K. Das Saha, A.K. Mandal,

Quercetin in vesicular delivery systems: Evaluation in combating arsenic-induced acute liver toxicity associated gene expression in rat model, Chem. Biol. Interact. 186 (2010) 61–71. doi:10.1016/j.cbi.2010.03.048.

[95]   F.-L. Yen, T.-H. Wu, L.-T. Lin, T.-M. Cham, C.-C. Lin, Naringenin-Loaded Nanoparticles Improve the Physicochemical Properties and the Hepatoprotective Effects of Naringenin in Orally-Administered Rats with CCl4-Induced Acute Liver Failure, Pharm. Res. 26 (2009) 893–902. doi:10.1007/s11095-008-9791-0.

[96]   J. Renugadevi, S.M. Prabu, Cadmium-induced hepatotoxicity in rats and the protective effect of naringenin, Exp. Toxicol. Pathol. 62 (2010) 171–181. doi:10.1016/j.etp.2009.03.010.

[97]   Y.-L. Zhao, G.-D. Zhou, H.-B. Yang, J.-B. Wang, L.-M. Shan, R. Li, X.-H. Xiao, Rhein protects against acetaminophen-induced hepatic and renal toxicity, Food Chem. Toxicol. 49 (2011) 1705–1710. doi:10.1016/j.fct.2011.04.011.

[98]   A. KoraMagazi, D. Wang, B. Yousef, M. Guerram, F. Yu, Rhein triggers apoptosis via induction of endoplasmic reticulum stress, caspase-4 and intracellular calcium in primary human hepatic HL-7702 cells, Biochem. Biophys. Res. Commun. 473 (2016) 230–236. doi:10.1016/j.bbrc.2016.03.084.

[99]   W.R. García-Niño, J. Pedraza-Chaverrí, Protective effect of curcumin against heavy metals-induced liver damage, Food Chem. Toxicol. 69 (2014) 182–201. doi:10.1016/j.fct.2014.04.016.

[100] L. Pari, R. Sivasankari, Effect of ellagic acid on cyclosporine A-induced oxidative damage in the liver of rats, Fundam. Clin. Pharmacol. 22 (2008) 395–401. doi:10.1111/j.1472-8206.2008.00609.x.

[101] X. Wan, M. Luo, X. Li, P. He, Hepatoprotective and anti-hepatocarcinogenic effects of glycyrrhizin and matrine, Chem. Biol. Interact. 181 (2009) 15–19. doi:10.1016/j.cbi.2009.04.013.

[102] X. Zhao, L. Wang, H. Zhang, D. Zhang, Z. Zhang, J. Zhang, Protective effect of artemisinin on chronic alcohol induced-liver damage in mice, Environ. Toxicol. Pharmacol. 52 (2017) 221–226. doi:10.1016/j.etap.2017.04.008.

[103] K. Czarnecka, N. Chufarova, K. Halczuk, K. Maciejewska, M. Girek, R. Skibiński, J.

Jończyk, M. Bajda, J. Kabziński, I. Majsterek, P. Szymański, Tetrahydroacridine derivatives with dichloronicotinic acid moiety as attractive, multipotent agents for Alzheimer's disease treatment, Eur. J. Med. Chem. 145 (2018) 760–769. doi:10.1016/j.ejmech.2018.01.014.

[104] T. Jung, Biodegradable nanoparticles for oral delivery of peptides: is there a role for polymers to affect mucosal uptake?, Eur. J. Pharm. Biopharm. 50 (2000) 147–160. doi:10.1016/S0939-6411(00)00084-9.

[105] A.P. Singh, A. Biswas, A. Shukla, P. Maiti, Targeted therapy in chronic diseases using nanomaterial-based drug delivery vehicles, Signal Transduct. Target. Ther. 4 (2019) 33. doi:10.1038/s41392-019-0068-3.

[106] I. Javed, G. Peng, Y. Xing, T. Yu, M. Zhao, A. Kakinen, A. Faridi, C.L. Parish, F. Ding, T.P. Davis, P.C. Ke, S. Lin, Inhibition of amyloid beta toxicity in zebrafish with a chaperone-gold nanoparticle dual strategy, Nat. Commun. 10 (2019) 3780. doi:10.1038/s41467-019-11762-0.

[107] P. Anand, A.B. Kunnumakkara, R.A. Newman, B.B. Aggarwal, Bioavailability of Curcumin: Problems and Promises, Mol. Pharm. 4 (2007) 807–818. doi:10.1021/mp700113r.

[108] A.J. Smith, P. Kavuru, L. Wojtas, M.J. Zaworotko, R.D. Shytle, Cocrystals of Quercetin with Improved Solubility and Oral Bioavailability, Mol. Pharm. 8 (2011) 1867–1876. doi:10.1021/mp200209j.

[109] A. Amri, J.C. Chaumeil, S. Sfar, C. Charrueau, Administration of resveratrol: What formulation solutions to bioavailability limitations?, J. Control. Release. 158 (2012) 182–193. doi:10.1016/j.jconrel.2011.09.083.

[110] B. Chellampillai, A.P. Pawar, Improved bioavailability of orally administered andrographolide from pH-sensitive nanoparticles, Eur. J. Drug Metab. Pharmacokinet. 35 (2011) 123–129. doi:10.1007/s13318-010-0016-7.

[111] I. Muqbil, A. Masood, F.H. Sarkar, R.M. Mohammad, A.S. Azmi, Progress in Nanotechnology Based Approaches to Enhance the Potential of Chemopreventive Agents, Cancers . 3 (2011). doi:10.3390/cancers3010428.

[112] H. Li, X. Zhao, Y. Ma, G. Zhai, L. Li, H. Lou, Enhancement of gastrointestinal

absorption of quercetin by solid lipid nanoparticles, J. Control. Release. 133 (2009) 238–244. doi:10.1016/j.jconrel.2008.10.002.

[113] M. Takahashi, S. Uechi, K. Takara, Y. Asikin, K. Wada, Evaluation of an Oral Carrier System in Rats: Bioavailability and Antioxidant Properties of Liposome-Encapsulated Curcumin, J. Agric. Food Chem. 57 (2009) 9141–9146. doi:10.1021/jf9013923.

[114] Y. Li, S. Sun, Q. Chang, L. Zhang, G. Wang, W. Chen, X. Miao, Y. Zheng, A Strategy for the Improvement of the Bioavailability and Antiosteoporosis Activity of BCS IV Flavonoid Glycosides through the Formulation of Their Lipophilic Aglycone into Nanocrystals, Mol. Pharm. 10 (2013) 2534–2542. doi:10.1021/mp300688t.

[115] J. Liang, Y. Liu, J. Liu, Z. Li, Q. Fan, Z. Jiang, F. Yan, Z. Wang, P. Huang, N. Feng, Chitosan-functionalized lipid-polymer hybrid nanoparticles for oral delivery of silymarin and enhanced lipid-lowering effect in NAFLD, J. Nanobiotechnology. 16 (2018) 64. doi:10.1186/s12951-018-0391-9.

[116] H. Luo, B. Jiang, B. Li, Z. Li, B.-H. Jiang, Y.C. Chen, Kaempferol nanoparticles achieve strong and selective inhibition of ovarian cancer cell viability, Int. J. Nanomedicine. 7 (2012) 3951–3959. doi:10.2147/IJN.S33670.

[117] I. Barwal, A. Sood, M. Sharma, B. Singh, S.C. Yadav, Development of stevioside Pluronic-F-68 copolymer based PLA-nanoparticles as an antidiabetic nanomedicine, Colloids Surfaces B Biointerfaces. 101 (2013) 510–516. doi:10.1016/j.colsurfb.2012.07.005.

[118] M. Roussaki, A. Gaitanarou, P.C. Diamanti, S. Vouyiouka, C. Papaspyrides, P. Kefalas, A. Detsi, Encapsulation of the natural antioxidant aureusidin in biodegradable PLA nanoparticles, Polym. Degrad. Stab. 108 (2014) 182–187. doi:10.1016/j.polymdegradstab.2014.08.004.

[119] L. Zou, W. Liu, W. Liu, R. Liang, T. Li, C. Liu, Y. Cao, J. Niu, Z. Liu, Characterization and Bioavailability of Tea Polyphenol Nanoliposome Prepared by Combining an Ethanol Injection Method with Dynamic High-Pressure Microfluidization, J. Agric. Food Chem. 62 (2014) 934–941. doi:10.1021/jf402886s.

[120] F.Y. Siu, S. Ye, H. Lin, S. Li, Galactosylated PLGA nanoparticles for the oral delivery of resveratrol: enhanced bioavailability and in vitro anti-inflammatory activity, Int. J.

Nanomedicine. 13 (2018) 4133–4144. doi:10.2147/IJN.S164235.

[121] Y. Ma, X. Zhao, J. Li, Q. Shen, The comparison of different daidzein-PLGA nanoparticles in increasing its oral bioavailability, Int. J. Nanomedicine. 7 (2012) 559–570. doi:10.2147/IJN.S27641.

[122] Y. Teow, S. Valiyaveettil, Active targeting of cancer cells using folic acid-conjugated platinum nanoparticles, Nanoscale. 2 (2010) 2607. doi:10.1039/c0nr00204f.

[123] C.-Q. You, H.-S. Wu, Z.-G. Gao, K. Sun, F.-H. Chen, W.A. Tao, B.-W. Sun, Subcellular co-delivery of two different site-oriented payloads based on multistage targeted polymeric nanoparticles for enhanced cancer therapy, J. Mater. Chem. B. 6 (2018) 6752–6766. doi:10.1039/C8TB02230E.

[124] Y. Luo, H. Yang, Y.-F. Zhou, B. Hu, Dual and multi-targeted nanoparticles for site-specific brain drug delivery, J. Control. Release. 317 (2020) 195–215. doi:10.1016/j.jconrel.2019.11.037.

[125] M. M. Yallapu, M. Jaggi, S. C. Chauhan, Curcumin Nanomedicine: A Road to Cancer Therapeutics, Curr. Pharm. Des. 19 (2013) 1994–2010. doi:10.2174/138161213805289219.

[126] M. Caldorera-Moore, N. Guimard, L. Shi, K. Roy, Designer nanoparticles: incorporating size, shape and triggered release into nanoscale drug carriers, Expert Opin. Drug Deliv. 7 (2010) 479–495. doi:10.1517/17425240903579971.

[127] I. Bala, V. Bhardwaj, S. Hariharan, S. V Kharade, N. Roy, M.N. V Ravi Kumar, Sustained release nanoparticulate formulation containing antioxidant-ellagic acid as potential prophylaxis system for oral administration, J. Drug Target. 14 (2006) 27–34. doi:10.1080/10611860600565987.

[128] J. Jacob, J.T. Haponiuk, S. Thomas, S. Gopi, Biopolymer based nanomaterials in drug delivery systems: A review, Mater. Today Chem. 9 (2018) 43–55. doi:https://doi.org/10.1016/j.mtchem.2018.05.002.

[129] B. Gibbens-Bandala, E. Morales-Avila, G. Ferro-Flores, C. Santos-Cuevas, L. Meléndez-Alafort, M. Trujillo-Nolasco, B. Ocampo-García, 177Lu-Bombesin-PLGA (paclitaxel): A targeted controlled-release nanomedicine for bimodal therapy of breast cancer, Mater. Sci. Eng. C. 105 (2019) 110043. doi:10.1016/j.msec.2019.110043.

[130] N.M. Khalil, T.C.F. do Nascimento, D.M. Casa, L.F. Dalmolin, A.C. de Mattos, I. Hoss, M.A. Romano, R.M. Mainardes, Pharmacokinetics of curcumin-loaded PLGA and PLGA–PEG blend nanoparticles after oral administration in rats, Colloids Surfaces B Biointerfaces. 101 (2013) 353–360. doi:10.1016/j.colsurfb.2012.06.024.

[131] X. Xie, Q. Tao, Y. Zou, F. Zhang, M. Guo, Y. Wang, H. Wang, Q. Zhou, S. Yu, PLGA Nanoparticles Improve the Oral Bioavailability of Curcumin in Rats: Characterizations and Mechanisms, J. Agric. Food Chem. 59 (2011) 9280–9289. doi:10.1021/jf202135j.

[132] D. Xiang, S. Shigdar, W. Yang, W. Duan, Q. Li, J. Lin, K. Liu, L. Li, Epithelial cell adhesion molecule aptamer functionalized PLGA-lecithin-curcumin-PEG nanoparticles for targeted drug delivery to human colorectal adenocarcinoma cells, Int. J. Nanomedicine. 9 (2014) 1083. doi:10.2147/IJN.S59779.

[133] K.R. Gajbhiye, V. Gajbhiye, I.A. Siddiqui, S. Pilla, V. Soni, Ascorbic acid tethered polymeric nanoparticles enable efficient brain delivery of galantamine: An in vitro-in vivo study, Sci. Rep. 7 (2017) 11086. doi:10.1038/s41598-017-11611-4.

[134] Y.-C. Kuo, R. Rajesh, Targeted delivery of rosmarinic acid across the blood–brain barrier for neuronal rescue using polyacrylamide-chitosan-poly(lactide- co -glycolide) nanoparticles with surface cross-reacting material 197 and apolipoprotein E, Int. J. Pharm. 528 (2017) 228–241. doi:10.1016/j.ijpharm.2017.05.039.

[135] A.K. Jain, K. Thanki, S. Jain, Co-encapsulation of Tamoxifen and Quercetin in Polymeric Nanoparticles: Implications on Oral Bioavailability, Antitumor Efficacy, and Drug-Induced Toxicity, Mol. Pharm. 10 (2013) 3459–3474. doi:10.1021/mp400311j.

[136] V. Kumar, A. Kumari, D. Kumar, S.K. Yadav, Biosurfactant stabilized anticancer biomolecule-loaded poly (d,l-lactide) nanoparticles, Colloids Surfaces B Biointerfaces. 117 (2014) 505–511. doi:10.1016/j.colsurfb.2014.01.057.

[137] C. Feng, X. Yuan, K. Chu, H. Zhang, W. Ji, M. Rui, Preparation and optimization of poly (lactic acid) nanoparticles loaded with fisetin to improve anti-cancer therapy, Int. J. Biol. Macromol. 125 (2019) 700–710. doi:10.1016/j.ijbiomac.2018.12.003.

[138] A. Kumari, S.K. Yadav, Y.B. Pakade, B. Singh, S.C. Yadav, Development of biodegradable nanoparticles for delivery of quercetin, Colloids Surfaces B

Biointerfaces. 80 (2010) 184–192. doi:10.1016/j.colsurfb.2010.06.002.

[139] A. Eatemadi, H. Daraee, H.T. Aiyelabegan, B. Negahdari, B. Rajeian, N. Zarghami, Synthesis and Characterization of Chrysin-loaded PCL-PEG-PCL nanoparticle and its effect on breast cancer cell line, Biomed. Pharmacother. 84 (2016) 1915–1922. doi:10.1016/j.biopha.2016.10.095.

[140] M. Zamani, M. Aghajanzadeh, K. Rostamizadeh, H. Kheiri Manjili, M. Fridoni, H. Danafar, In vivo study of poly (ethylene glycol)-poly (caprolactone)-modified folic acid nanocarriers as a pH responsive system for tumor-targeted co-delivery of tamoxifen and quercetin, J. Drug Deliv. Sci. Technol. 54 (2019) 101283. doi:10.1016/j.jddst.2019.101283.

[141] C. Moorthi, K. Kathiresan, Curcumin–Piperine/Curcumin–Quercetin/Curcumin–Silibinin dual drug-loaded nanoparticulate combination therapy: A novel approach to target and treat multidrug-resistant cancers, J. Med. Hypotheses Ideas. 7 (2013) 15–20. doi:10.1016/j.jmhi.2012.10.005.

[142] P. Mukhopadhyay, S. Maity, S. Mandal, A.S. Chakraborti, A.K. Prajapati, P.P. Kundu, Preparation, characterization and in vivo evaluation of pH sensitive, safe quercetin-succinylated chitosan-alginate core-shell-corona nanoparticle for diabetes treatment, Carbohydr. Polym. 182 (2018) 42–51. doi:10.1016/j.carbpol.2017.10.098.

[143] X. Huang, Y. Dai, J. Cai, N. Zhong, H. Xiao, D.J. McClements, K. Hu, Resveratrol encapsulation in core-shell biopolymer nanoparticles: Impact on antioxidant and anticancer activities, Food Hydrocoll. 64 (2017) 157–165. doi:10.1016/j.foodhyd.2016.10.029.

[144] J.-G. Leu, S.-A. Chen, H.-M. Chen, W.-M. Wu, C.-F. Hung, Y.-D. Yao, C.-S. Tu, Y.-J. Liang, The effects of gold nanoparticles in wound healing with antioxidant epigallocatechin gallate and α-lipoic acid, Nanomedicine Nanotechnology, Biol. Med. 8 (2012) 767–775. doi:10.1016/j.nano.2011.08.013.

[145] F.G. Milanezi, L.M. Meireles, M.M. de Christo Scherer, J.P. de Oliveira, A.R. da Silva, M.L. de Araujo, D.C. Endringer, M. Fronza, M.C.C. Guimarães, R. Scherer, Antioxidant, antimicrobial and cytotoxic activities of gold nanoparticles capped with quercetin, Saudi Pharm. J. 27 (2019) 968–974. doi:10.1016/j.jsps.2019.07.005.

[146] J. Gubitosa, V. Rizzi, P. Fini, R. Del Sole, A. Lopedota, V. Laquintana, N. Denora, A. Agostiano, P. Cosma, Multifunctional green synthetized gold nanoparticles/chitosan/ellagic acid self-assembly: Antioxidant, sun filter and tyrosinase-inhibitor properties, Mater. Sci. Eng. C. 106 (2020) 110170. doi:10.1016/j.msec.2019.110170.

[147] G.H. Lee, S.J. Lee, S.W. Jeong, H.-C. Kim, G.Y. Park, S.G. Lee, J.H. Choi, Antioxidative and antiinflammatory activities of quercetin-loaded silica nanoparticles, Colloids Surfaces B Biointerfaces. 143 (2016) 511–517. doi:10.1016/j.colsurfb.2016.03.060.

[148] M.J. Kim, S.U. Rehman, F.U. Amin, M.O. Kim, Enhanced neuroprotection of anthocyanin-loaded PEG-gold nanoparticles against Aβ1-42-induced neuroinflammation and neurodegeneration via the NF-KB /JNK/GSK3β signaling pathway, Nanomedicine Nanotechnology, Biol. Med. 13 (2017) 2533–2544. doi:10.1016/j.nano.2017.06.022.

[149] R. Nosratabadi, M. Rastin, M. Sankian, D. Haghmorad, M. Mahmoudi, Hyperforin-loaded gold nanoparticle alleviates experimental autoimmune encephalomyelitis by suppressing Th1 and Th17 cells and upregulating regulatory T cells, Nanomedicine Nanotechnology, Biol. Med. 12 (2016) 1961–1971. doi:10.1016/j.nano.2016.04.001.

[150] D.-Y. Kim, M. Kim, S. Shinde, J.-S. Sung, G. Ghodake, Cytotoxicity and antibacterial assessment of gallic acid capped gold nanoparticles, Colloids Surfaces B Biointerfaces. 149 (2017) 162–167. doi:10.1016/j.colsurfb.2016.10.017.

[151] G.A. Islan, S. Das, M.L. Cacicedo, A. Halder, A. Mukherjee, M.L. Cuestas, P. Roy, G.R. Castro, A. Mukherjee, Silybin-conjugated gold nanoparticles for antimicrobial chemotherapy against Gram-negative bacteria, J. Drug Deliv. Sci. Technol. 53 (2019) 101181. doi:10.1016/j.jddst.2019.101181.

[152] Á. Artiga, I. Serrano-Sevilla, L. De Matteis, S.G. Mitchell, J.M. de la Fuente, Current status and future perspectives of gold nanoparticle vectors for siRNA delivery, J. Mater. Chem. B. 7 (2019) 876–896. doi:10.1039/C8TB02484G.

[153] Y. Lyu, M. Yu, Q. Liu, Q. Zhang, Z. Liu, Y. Tian, D. Li, M. Changdao, Synthesis of silver nanoparticles using oxidized amylose and combination with curcumin for enhanced antibacterial activity, Carbohydr. Polym. 230 (2020) 115573.

doi:10.1016/j.carbpol.2019.115573.

[154] S.N. Sunil Gowda, S. Rajasowmiya, V. Vadivel, S. Banu Devi, A. Celestin Jerald, S. Marimuthu, N. Devipriya, Gallic acid-coated sliver nanoparticle alters the expression of radiation-induced epithelial-mesenchymal transition in non-small lung cancer cells, Toxicol. Vitr. 52 (2018) 170–177. doi:10.1016/j.tiv.2018.06.015.

[155] K. Ranoszek-Soliwoda, E. Tomaszewska, K. Małek, G. Celichowski, P. Orlowski, M. Krzyzowska, J. Grobelny, The synthesis of monodisperse silver nanoparticles with plant extracts, Colloids Surfaces B Biointerfaces. 177 (2019) 19–24. doi:10.1016/j.colsurfb.2019.01.037.

[156] E.-Y. Ahn, H. Jin, Y. Park, Assessing the antioxidant, cytotoxic, apoptotic and wound healing properties of silver nanoparticles green-synthesized by plant extracts, Mater. Sci. Eng. C. 101 (2019) 204–216. doi:10.1016/j.msec.2019.03.095.

[157] A. Halder, S. Das, D. Ojha, D. Chattopadhyay, A. Mukherjee, Highly monodispersed gold nanoparticles synthesis and inhibition of herpes simplex virus infections, Mater. Sci. Eng. C. 89 (2018) 413–421. doi:10.1016/j.msec.2018.04.005.

[158] I. Lynch, K.A. Dawson, Protein-nanoparticle interactions, Nano Today. 3 (2008) 40–47. doi:10.1016/S1748-0132(08)70014-8.

[159] D. Cassano, A.-K. Mapanao, M. Summa, Y. Vlamidis, G. Giannone, M. Santi, E. Guzzolino, L. Pitto, L. Poliseno, R. Bertorelli, V. Voliani, Biosafety and Biokinetics of Noble Metals: The Impact of Their Chemical Nature, ACS Appl. Bio Mater. 2 (2019) 4464–4470. doi:10.1021/acsabm.9b00630.

[160] S.-T. Yang, Y. Liu, Y.-W. Wang, A. Cao, Biosafety and Bioapplication of Nanomaterials by Designing Protein-Nanoparticle Interactions, Small. 9 (2013) 1635–1653. doi:10.1002/smll.201201492.

[161] A. Frydman-Marom, M. Rechter, I. Shefler, Y. Bram, D.E. Shalev, E. Gazit, Cognitive-Performance Recovery of Alzheimer's Disease Model Mice by Modulation of Early Soluble Amyloidal Assemblies, Angew. Chemie Int. Ed. 48 (2009) 1981–1986. doi:10.1002/anie.200802123.

[162] T.M. Marcinko, J. Dong, R. LeBlanc, K. V Daborowski, R.W. Vachet, Small molecule-mediated inhibition of β-2-microglobulin-based amyloid fibril formation, J.

Biol. Chem. . 292 (2017) 10630–10638. doi:10.1074/jbc.M116.774083.

[163] J. Pujols, S. Peña-Díaz, D.F. Lázaro, F. Peccati, F. Pinheiro, D. González, A. Carija, S. Navarro, M. Conde-Giménez, J. García, S. Guardiola, E. Giralt, X. Salvatella, J. Sancho, M. Sodupe, T.F. Outeiro, E. Dalfó, S. Ventura, Small molecule inhibits α-synuclein aggregation, disrupts amyloid fibrils, and prevents degeneration of dopaminergic neurons, Proc. Natl. Acad. Sci. 115 (2018) 10481 LP – 10486. doi:10.1073/pnas.1804198115.

[164] K. Ono, K. Hasegawa, H. Naiki, M. Yamada, Curcumin has potent anti-amyloidogenic effects for Alzheimer's β-amyloid fibrils in vitro, J. Neurosci. Res. 75 (2004) 742–750. doi:10.1002/jnr.20025.

[165] Z. Wang, X. Dong, Y. Sun, Mixed Carboxyl and Hydrophobic Dendrimer Surface Inhibits Amyloid-β Fibrillation: New Insight from the Generation Number Effect, Langmuir. 35 (2019) 14681–14687. doi:10.1021/acs.langmuir.9b02527.

[166] H. Skaat, G. Shafir, S. Margel, Acceleration and inhibition of amyloid-β fibril formation by peptide-conjugated fluorescent-maghemite nanoparticles, J. Nanoparticle Res. 13 (2011) 3521–3534. doi:10.1007/s11051-011-0276-4.

[167] F.U. Hartl, A. Bracher, M. Hayer-Hartl, Molecular chaperones in protein folding and proteostasis, Nature. 475 (2011) 324–332. doi:10.1038/nature10317.

[168] H.M. Sanders, B. Jovcevski, J.A. Carver, T.L. Pukala, The molecular chaperone β-casein prevents amorphous and fibrillar aggregation of α-lactalbumin by stabilisation of dynamic disorder, Biochem. J. 477 (2020) 629–643. doi:10.1042/BCJ20190638.

[169] R. Montserret, M.J. McLeish, A. Böckmann, C. Geourjon, F. Penin, Involvement of Electrostatic Interactions in the Mechanism of Peptide Folding Induced by Sodium Dodecyl Sulfate Binding, Biochemistry. 39 (2000) 8362–8373. doi:10.1021/bi000208x.

[170] S.S.-S. Wang, Y.-T. Chen, S.-W. Chou, Inhibition of amyloid fibril formation of β-amyloid peptides via the amphiphilic surfactants, Biochim. Biophys. Acta - Mol. Basis Dis. 1741 (2005) 307–313. doi:10.1016/j.bbadis.2005.05.004.

[171] S. Kazmi, A.A. Mujeeb, M. Owais, Cyclic undecapeptide Cyclosporin A mediated inhibition of amyloid synthesis: Implications in alleviation of amyloid induced

neurotoxicity, Sci. Rep. 8 (2018) 17283. doi:10.1038/s41598-018-35645-4.

[172] A. Lorenzo, B.A. Yankner, Beta-amyloid neurotoxicity requires fibril formation and is inhibited by congo red, Proc. Natl. Acad. Sci. 91 (1994) 12243 LP – 12247. doi:10.1073/pnas.91.25.12243.

[173] S. Sinha, D.H.J. Lopes, Z. Du, E.S. Pang, A. Shanmugam, A. Lomakin, P. Talbiersky, A. Tennstaedt, K. McDaniel, R. Bakshi, P.-Y. Kuo, M. Ehrmann, G.B. Benedek, J.A. Loo, F.-G. Klärner, T. Schrader, C. Wang, G. Bitan, Lysine-Specific Molecular Tweezers Are Broad-Spectrum Inhibitors of Assembly and Toxicity of Amyloid Proteins, J. Am. Chem. Soc. 133 (2011) 16958–16969. doi:10.1021/ja206279b.

[174] R. Zhang, C. Zhang, Q. Zhao, D. Li, Spectrin: Structure, function and disease, Sci. China Life Sci. 56 (2013) 1076–1085. doi:10.1007/s11427-013-4575-0.

[175] R.S. Sankhala, M.J. Swamy, The Major Protein of Bovine Seminal Plasma, PDC-109, Is a Molecular Chaperone, Biochemistry. 49 (2010) 3908–3918. doi:10.1021/bi100051d.

[176] K. Ikeda, T. Okada, S. Sawada, K. Akiyoshi, K. Matsuzaki, Inhibition of the formation of amyloid β-protein fibrils using biocompatible nanogels as artificial chaperones, FEBS Lett. 580 (2006) 6587–6595. doi:10.1016/j.febslet.2006.11.009.

[177] T.S. Jarvela, H.A. Lam, M. Helwig, N. Lorenzen, D.E. Otzen, P.J. McLean, N.T. Maidment, I. Lindberg, The neural chaperone proSAAS blocks α-synuclein fibrillation and neurotoxicity, Proc. Natl. Acad. Sci. 113 (2016) E4708 LP-E4715. doi:10.1073/pnas.1601091113.

[178] S. Movaghati, A.A. Moosavi-Movahedi, F. Khodagholi, H. Digaleh, E. Kachooei, N. Sheibani, Sodium dodecyl sulphate modulates the fibrillation of human serum albumin in a dose-dependent manner and impacts the PC12 cells retraction, Colloids Surfaces B Biointerfaces. 122 (2014) 341–349. doi:10.1016/j.colsurfb.2014.07.002.

[179] Z. Yaseen, S.U. Rehman, M. Tabish, A.H. Shalla, Kabir-ud-Din, Modulation of bovine serum albumin fibrillation by ester bonded and conventional gemini surfactants, RSC Adv. 5 (2015) 58616–58624. doi:10.1039/C5RA08923A.

[180] M. Akram, I.A. Bhat, Kabir-ud-Din, New insights into binding interaction of novel ester-functionalized m-E2-m gemini surfactants with lysozyme: a detailed

multidimensional study, RSC Adv. 5 (2015) 102780–102794.
doi:10.1039/C5RA20576J.

[181] E.K. Kumar, N.P. Prabhu, Differential effects of ionic and non-ionic surfactants on lysozyme fibrillation, Phys. Chem. Chem. Phys. 16 (2014) 24076–24088. doi:10.1039/C4CP02423K.

[182] N. Sharma, R. Rajan, S. Makhaik, K. Matsumura, Comparative Study of Protein Aggregation Arrest by Zwitterionic Polysulfobetaines: Using Contrasting Raft Agents, ACS Omega. 4 (2019) 12186—12193. doi:10.1021/acsomega.9b01409.

[183] M. Zaman, S.M. Zakariya, S. Nusrat, T.I. Chandel, S.M. Meeran, M.R. Ajmal, P. Alam, Wahiduzzaman, R.H. Khan, Cysteine as a potential anti-amyloidogenic agent with protective ability against amyloid induced cytotoxicity, Int. J. Biol. Macromol. 105 (2017) 556–565. doi:10.1016/j.ijbiomac.2017.07.083.

[184] M.M. Varughese, J. Newman, Inhibitory Effects of Arginine on the Aggregation of Bovine Insulin, J. Biophys. 2012 (2012) 434289. doi:10.1155/2012/434289.

[185] E. Takai, K. Uda, T. Yoshida, T. Zako, M. Maeda, K. Shiraki, Cysteine inhibits the fibrillisation and cytotoxicity of amyloid-β 40 and 42: implications for the contribution of the thiophilic interaction, Phys. Chem. Chem. Phys. 16 (2014) 3566–3572. doi:10.1039/C3CP54245A.

[186] J. Kumar, V. Sim, D-amino acid-based peptide inhibitors as early or preventative therapy in Alzheimer disease, Prion. 8 (2014) 119–124. doi:10.4161/pri.28220.

[187] T. KAWASAKI, K. ONODERA, S. KAMIJO, Selection of Peptide Inhibitors of Soluble Aβ1-42 Oligomer Formation by Phage Display, Biosci. Biotechnol. Biochem. 74 (2010) 2214–2219. doi:10.1271/bbb.100388.

[188] T. Takahashi, H. Mihara, Peptide and Protein Mimetics Inhibiting Amyloid β-Peptide Aggregation, Acc. Chem. Res. 41 (2008) 1309–1318. doi:10.1021/ar8000475.

[189] B. Dorgeret, L. Khemtémourian, I. Correia, J.-L. Soulier, O. Lequin, S. Ongeri, Sugar-based peptidomimetics inhibit amyloid β-peptide aggregation, Eur. J. Med. Chem. 46 (2011) 5959–5969. doi:10.1016/j.ejmech.2011.10.008.

[190] K.M. Pate, B.J. Kim, E. V Shusta, R.M. Murphy, Transthyretin Mimetics as Anti-β-Amyloid Agents: A Comparison of Peptide and Protein Approaches, ChemMedChem.

13 (2018) 968–979. doi:10.1002/cmdc.201800031.

[191] A.R. Salomon, K.J. Marcinowski, R.P. Friedland, M.G. Zagorski, Nicotine Inhibits Amyloid Formation by the β-Peptide, Biochemistry. 35 (1996) 13568–13578. doi:10.1021/bi9617264.

[192] D. Howlett, P. Cutler, S. Heales, P. Camilleri, Hemin and related porphyrins inhibit β-amyloid aggregation, FEBS Lett. 417 (1997) 249–251. doi:10.1016/S0014-5793(97)01290-8.

[193] T. Tomiyama, A. Shoji, K. Kataoka, Y. Suwa, S. Asano, H. Kaneko, N. Endo, Inhibition of Amyloid Protein Aggregation and Neurotoxicity by Rifampicin: ITS POSSIBLE FUNCTION AS A HYDROXYL RADICAL SCAVENGER , J. Biol. Chem. . 271 (1996) 6839–6844. doi:10.1074/jbc.271.12.6839.

[194] M. Pappolla, P. Bozner, C. Soto, H. Shao, N.K. Robakis, M. Zagorski, B. Frangione, J. Ghiso, Inhibition of Alzheimer β-Fibrillogenesis by Melatonin, J. Biol. Chem. . 273 (1998) 7185–7188. doi:10.1074/jbc.273.13.7185.

[195] T. Thomas, T.G. Nadackal, K. Thomas, Aspirin and non-steroidal anti-inflammatory drugs inhibit amyloid-β aggregation, Neuroreport. 12 (2001) 3263–3267. doi:10.1097/00001756-200110290-00024.

[196] Y. Sun, A. Kakinen, C. Zhang, Y. Yang, A. Faridi, T.P. Davis, W. Cao, P.C. Ke, F. Ding, Amphiphilic surface chemistry of fullerenols is necessary for inhibiting the amyloid aggregation of alpha-synuclein NACore, Nanoscale. 11 (2019) 11933–11945. doi:10.1039/C9NR02407G.

[197] T. Kukar, T.E. Golde, Possible mechanisms of action of NSAIDs and related compounds that modulate gamma-secretase cleavage, Curr. Top. Med. Chem. 8 (2008) 47–53. doi:10.2174/156802608783334042.

[198] J.I.E. LI, M.I.N. ZHU, A.M.Y.B. MANNING-BOG, D.A. DI MONTE, A.L. FINK, Dopamine and L-dopa disaggregate amyloid fibrils: implications for Parkinson's and Alzheimer's disease, FASEB J. 18 (2004) 962–964. doi:10.1096/fj.03-0770fje.

[199] M.R. Ajmal, T.I. Chandel, P. Alam, N. Zaidi, M. Zaman, S. Nusrat, M.V. Khan, M.K. Siddiqi, Y.E. Shahein, M.H. Mahmoud, G. Badr, R.H. Khan, Fibrillogenesis of human serum albumin in the presence of levodopa- spectroscopic, calorimetric and

microscopic studies, Int. J. Biol. Macromol. 94 (2017) 301–308.
doi:10.1016/j.ijbiomac.2016.10.025.

[200] C.A. Braga, C. Follmer, F.L. Palhano, E. Khattar, M.S. Freitas, L. Romão, S. Di
Giovanni, H.A. Lashuel, J.L. Silva, D. Foguel, The Anti-Parkinsonian Drug Selegiline
Delays the Nucleation Phase of α-Synuclein Aggregation Leading to the Formation of
Nontoxic Species, J. Mol. Biol. 405 (2011) 254–273. doi:10.1016/j.jmb.2010.10.027.

[201] M.K. Siddiqi, P. Alam, S.K. Chaturvedi, M.V. Khan, S. Nusrat, S. Malik, R.H. Khan,
Capreomycin inhibits the initiation of amyloid fibrillation and suppresses amyloid
induced cell toxicity, Biochim. Biophys. Acta - Proteins Proteomics. 1866 (2018) 549–
557. doi:10.1016/j.bbapap.2018.02.005.

[202] M. Lundqvist, J. Stigler, T. Cedervall, T. Berggård, M.B. Flanagan, I. Lynch, G. Elia,
K. Dawson, The evolution of the protein corona around nanoparticles: A test study,
ACS Nano. 5 (2011) 7503–7509. doi:10.1021/nn202458g.

[203] M. Lundqvist, J. Stigler, G. Elia, I. Lynch, T. Cedervall, K.A. Dawson, Nanoparticle
size and surface properties determine the protein corona with possible implications for
biological impacts, Proc. Natl. Acad. Sci. 105 (2008) 14265–14270.
doi:10.1073/pnas.0805135105.

[204] M.J. Roberti, M. Morgan, G. Menéndez, L.I. Pietrasanta, T.M. Jovin, E.A. Jares-
Erijman, Quantum Dots As Ultrasensitive Nanoactuators and Sensors of Amyloid
Aggregation in Live Cells, J. Am. Chem. Soc. 131 (2009) 8102–8107.
doi:10.1021/ja900225w.

[205] M. Mahmoudi, H.R. Kalhor, S. Laurent, I. Lynch, Protein fibrillation and nanoparticle
interactions: opportunities and challenges, Nanoscale. 5 (2013) 2570.
doi:10.1039/c3nr33193h.

[206] C. Cabaleiro-Lago, F. Quinlan-Pluck, I. Lynch, K.A. Dawson, S. Linse, Dual Effect of
Amino Modified Polystyrene Nanoparticles on Amyloid β Protein Fibrillation, ACS
Chem. Neurosci. 1 (2010) 279–287. doi:10.1021/cn900027u.

[207] M.P. Monopoli, D. Walczyk, A. Campbell, G. Elia, I. Lynch, F. Baldelli Bombelli,
K.A. Dawson, Physical–Chemical Aspects of Protein Corona: Relevance to in Vitro
and in Vivo Biological Impacts of Nanoparticles, J. Am. Chem. Soc. 133 (2011) 2525–

2534. doi:10.1021/ja107583h.

[208] S.-Y. Hung, W.-M. Fu, Drug candidates in clinical trials for Alzheimer's disease, J. Biomed. Sci. 24 (2017) 47. doi:10.1186/s12929-017-0355-7.

[209] S. Salloway, R. Sperling, R. Keren, A.P. Porsteinsson, C.H. van Dyck, P.N. Tariot, S. Gilman, D. Arnold, S. Abushakra, C. Hernandez, G. Crans, E. Liang, G. Quinn, M. Bairu, A. Pastrak, J.M. Cedarbaum, A phase 2 randomized trial of ELND005, scyllo-inositol, in mild to moderate Alzheimer disease, Neurology. 77 (2011) 1253 LP – 1262. doi:10.1212/WNL.0b013e3182309fa5.

[210] F. van Bebber, D. Paquet, A. Hruscha, B. Schmid, C. Haass, Methylene blue fails to inhibit Tau and polyglutamine protein dependent toxicity in zebrafish, Neurobiol. Dis. 39 (2010) 265–271. doi:10.1016/j.nbd.2010.03.023.

[211] M.G. Savelieff, A.S. DeToma, J.S. Derrick, M.H. Lim, The Ongoing Search for Small Molecules to Study Metal-Associated Amyloid-β Species in Alzheimer's Disease, Acc. Chem. Res. 47 (2014) 2475–2482. doi:10.1021/ar500152x.

[212] J. Burns, T. Yokota, H. Ashihara, M.E.J. Lean, A. Crozier, Plant Foods and Herbal Sources of Resveratrol, J. Agric. Food Chem. 50 (2002) 3337–3340. doi:10.1021/jf0112973.

[213] B. Sultana, F. Anwar, Flavonols (kaempferol, quercetin, myricetin) contents of selected fruits, vegetables and medicinal plants, Food Chem. 108 (2008) 879–884. doi:10.1016/j.foodchem.2007.11.053.

[214] L.T. Abe, F.M. Lajolo, M.I. Genovese, Potential dietary sources of ellagic acid and other antioxidants among fruits consumed in Brazil: Jabuticaba (Myrciaria jaboticaba (Vell.) Berg), J. Sci. Food Agric. 92 (2012) 1679–1687. doi:10.1002/jsfa.5531.

[215] N.A. Fazili, A. Naeem, Anti-fibrillation potency of caffeic acid against an antidepressant induced fibrillogenesis of human α-synuclein: Implications for Parkinson's disease, Biochimie. 108 (2015) 178–185. doi:10.1016/j.biochi.2014.11.011.

[216] A. Cornejo, F. Aguilar Sandoval, L. Caballero, L. Machuca, P. Muñoz, J. Caballero, G. Perry, A. Ardiles, C. Areche, F. Melo, Rosmarinic acid prevents fibrillization and diminishes vibrational modes associated to β sheet in tau protein linked to Alzheimer's

disease, J. Enzyme Inhib. Med. Chem. 32 (2017) 945–953.
doi:10.1080/14756366.2017.1347783.

[217] B. Ren, Y. Liu, Y. Zhang, Y. Cai, X. Gong, Y. Chang, L. Xu, J. Zheng, Genistein: A Dual Inhibitor of Both Amyloid β and Human Islet Amylin Peptides, ACS Chem. Neurosci. 9 (2018) 1215–1224. doi:10.1021/acschemneuro.8b00039.

[218] M. Sato, K. Murakami, M. Uno, Y. Nakagawa, S. Katayama, K. Akagi, Y. Masuda, K. Takegoshi, K. Irie, Site-specific Inhibitory Mechanism for Amyloid β42 Aggregation by Catechol-type Flavonoids Targeting the Lys Residues, J. Biol. Chem. 288 (2013) 23212–23224. doi:10.1074/jbc.M113.464222.

[219] F.L. Palhano, J. Lee, N.P. Grimster, J.W. Kelly, Toward the Molecular Mechanism(s) by Which EGCG Treatment Remodels Mature Amyloid Fibrils, J. Am. Chem. Soc. 135 (2013) 7503–7510. doi:10.1021/ja3115696.

[220] T. Ishii, T. Mori, T. Tanaka, D. Mizuno, R. Yamaji, S. Kumazawa, T. Nakayama, M. Akagawa, Covalent modification of proteins by green tea polyphenol (–)-epigallocatechin-3-gallate through autoxidation, Free Radic. Biol. Med. 45 (2008) 1384–1394. doi:10.1016/j.freeradbiomed.2008.07.023.

[221] P. Cao, D.P. Raleigh, Analysis of the Inhibition and Remodeling of Islet Amyloid Polypeptide Amyloid Fibers by Flavanols, Biochemistry. 51 (2012) 2670–2683. doi:10.1021/bi2015162.

[222] L.-H. Tu, L.M. Young, A.G. Wong, A.E. Ashcroft, S.E. Radford, D.P. Raleigh, Mutational Analysis of the Ability of Resveratrol To Inhibit Amyloid Formation by Islet Amyloid Polypeptide: Critical Evaluation of the Importance of Aromatic–Inhibitor and Histidine–Inhibitor Interactions, Biochemistry. 54 (2015) 666–676. doi:10.1021/bi501016r.

[223] B. Cheng, H. Gong, H. Xiao, R.B. Petersen, L. Zheng, K. Huang, Inhibiting toxic aggregation of amyloidogenic proteins: A therapeutic strategy for protein misfolding diseases, Biochim. Biophys. Acta - Gen. Subj. 1830 (2013) 4860–4871. doi:10.1016/j.bbagen.2013.06.029.

[224] W.M. Berhanu, A.E. Masunov, Atomistic mechanism of polyphenol amyloid aggregation inhibitors: molecular dynamics study of Curcumin, Exifone, and

Myricetin interaction with the segment of tau peptide oligomer, J. Biomol. Struct. Dyn. 33 (2015) 1399–1411. doi:10.1080/07391102.2014.951689.

[225] P. Nedumpully-Govindan, A. Kakinen, E.H. Pilkington, T.P. Davis, P. Chun Ke, F. Ding, Stabilizing Off-pathway Oligomers by Polyphenol Nanoassemblies for IAPP Aggregation Inhibition, Sci. Rep. 6 (2016) 19463. doi:10.1038/srep19463.

[226] K.M. Nelson, J.L. Dahlin, J. Bisson, J. Graham, G.F. Pauli, M.A. Walters, The Essential Medicinal Chemistry of Curcumin, J. Med. Chem. 60 (2017) 1620–1637. doi:10.1021/acs.jmedchem.6b00975.

[227] M. Landau, M.R. Sawaya, K.F. Faull, A. Laganowsky, L. Jiang, S.A. Sievers, J. Liu, J.R. Barrio, D. Eisenberg, Towards a Pharmacophore for Amyloid, PLoS Biol. 9 (2011) e1001080. doi:10.1371/journal.pbio.1001080.

[228] J. Bieschke, J. Russ, R.P. Friedrich, D.E. Ehrnhoefer, H. Wobst, K. Neugebauer, E.E. Wanker, EGCG remodels mature alpha-synuclein and amyloid-beta fibrils and reduces cellular toxicity, Proc. Natl. Acad. Sci. U. S. A. 107 (2010) 7710–7715. doi:10.1073/pnas.0910723107.

[229] M.F.M. Engel, C.C. vandenAkker, M. Schleeger, K.P. Velikov, G.H. Koenderink, M. Bonn, The Polyphenol EGCG Inhibits Amyloid Formation Less Efficiently at Phospholipid Interfaces than in Bulk Solution, J. Am. Chem. Soc. 134 (2012) 14781–14788. doi:10.1021/ja3031664.

[230] N. Popovych, J.R. Brender, R. Soong, S. Vivekanandan, K. Hartman, V. Basrur, P.M. Macdonald, A. Ramamoorthy, Site Specific Interaction of the Polyphenol EGCG with the SEVI Amyloid Precursor Peptide PAP(248–286), J. Phys. Chem. B. 116 (2012) 3650–3658. doi:10.1021/jp2121577.

[231] A. Thapa, S.D. Jett, E.Y. Chi, Curcumin Attenuates Amyloid-β Aggregate Toxicity and Modulates Amyloid-β Aggregation Pathway, ACS Chem. Neurosci. 7 (2016) 56–68. doi:10.1021/acschemneuro.5b00214.

[232] N.N. Jha, D. Ghosh, S. Das, A. Anoop, R.S. Jacob, P.K. Singh, N. Ayyagari, I.N.N. Namboothiri, S.K. Maji, Effect of curcumin analogs onα-synuclein aggregation and cytotoxicity, Sci. Rep. 6 (2016) 28511. doi:10.1038/srep28511.

[233] D.E. Ehrnhoefer, J. Bieschke, A. Boeddrich, M. Herbst, L. Masino, R. Lurz, S.

Engemann, A. Pastore, E.E. Wanker, EGCG redirects amyloidogenic polypeptides into unstructured, off-pathway oligomers, Nat. Struct. Mol. Biol. 15 (2008) 558–566. doi:10.1038/nsmb.1437.

[234] G. Liu, J.C. Gaines, K.J. Robbins, N.D. Lazo, Kinetic Profile of Amyloid Formation in the Presence of an Aromatic Inhibitor by Nuclear Magnetic Resonance, ACS Med. Chem. Lett. 3 (2012) 856–859. doi:10.1021/ml300147m.

[235] R.C. George, J. Lew, D.J. Graves, Interaction of Cinnamaldehyde and Epicatechin with Tau: Implications of Beneficial Effects in Modulating Alzheimer's Disease Pathogenesis, J. Alzheimer's Dis. 36 (2013) 21–40. doi:10.3233/JAD-122113.

[236] W. Li, J.B. Sperry, A. Crowe, J.Q. Trojanowski, A.B. Smith III, V.M.-Y. Lee, Inhibition of tau fibrillization by oleocanthal via reaction with the amino groups of tau, J. Neurochem. 110 (2009) 1339–1351. doi:10.1111/j.1471-4159.2009.06224.x.

[237] P.K. Singh, V. Kotia, D. Ghosh, G.M. Mohite, A. Kumar, S.K. Maji, Curcumin Modulates α-Synuclein Aggregation and Toxicity, ACS Chem. Neurosci. 4 (2013) 393–407. doi:10.1021/cn3001203.

[238] J.M. Lopez del Amo, U. Fink, M. Dasari, G. Grelle, E.E. Wanker, J. Bieschke, B. Reif, Structural Properties of EGCG-Induced, Nontoxic Alzheimer's Disease Aβ Oligomers, J. Mol. Biol. 421 (2012) 517–524. doi:10.1016/j.jmb.2012.01.013.

[239] A.R.A. Ladiwala, J.C. Lin, S.S. Bale, A.M. Marcelino-Cruz, M. Bhattacharya, J.S. Dordick, P.M. Tessier, Resveratrol Selectively Remodels Soluble Oligomers and Fibrils of Amyloid Aβ into Off-pathway Conformers, J. Biol. Chem. . 285 (2010) 24228–24237. doi:10.1074/jbc.M110.133108.

[240] A.R.A. Ladiwala, J.S. Dordick, P.M. Tessier, Aromatic Small Molecules Remodel Toxic Soluble Oligomers of Amyloid β through Three Independent Pathways, J. Biol. Chem. . 286 (2011) 3209–3218. doi:10.1074/jbc.M110.173856.

[241] F. Chiti, C.M. Dobson, Protein Misfolding, Functional Amyloid, and Human Disease, Annu. Rev. Biochem. 75 (2006) 333–366. doi:10.1146/annurev.biochem.75.101304.123901.

[242] R. Kayed, E. Head, J.L. Thompson, T.M. McIntire, S.C. Milton, C.W. Cotman, C.G. Glabe, Common Structure of Soluble Amyloid Oligomers Implies Common

Mechanism of Pathogenesis, Science (80-.). 300 (2003) 486 LP – 489. doi:10.1126/science.1079469.

[243] C.G. Glabe, Structural Classification of Toxic Amyloid Oligomers, J. Biol. Chem. . 283 (2008) 29639–29643. doi:10.1074/jbc.R800016200.

[244] D.-P. Hong, A.L. Fink, V.N. Uversky, Structural characteristics of alpha-synuclein oligomers stabilized by the flavonoid baicalein, J. Mol. Biol. 383 (2008) 214–223. doi:10.1016/j.jmb.2008.08.039.

[245] M. Zhu, S. Rajamani, J. Kaylor, S. Han, F. Zhou, A.L. Fink, The Flavonoid Baicalein Inhibits Fibrillation of α-Synuclein and Disaggregates Existing Fibrils, J. Biol. Chem. . 279 (2004) 26846–26857. doi:10.1074/jbc.M403129200.

[246] S. Taniguchi, N. Suzuki, M. Masuda, S. Hisanaga, T. Iwatsubo, M. Goedert, M. Hasegawa, Inhibition of Heparin-induced Tau Filament Formation by Phenothiazines, Polyphenols, and Porphyrins, J. Biol. Chem. . 280 (2005) 7614–7623. doi:10.1074/jbc.M408714200.

[247] M.S. Planchard, M.A. Samel, A. Kumar, V. Rangachari, The Natural Product Betulinic Acid Rapidly Promotes Amyloid-β Fibril Formation at the Expense of Soluble Oligomers, ACS Chem. Neurosci. 3 (2012) 900–908. doi:10.1021/cn300030a.

[248] D.S. Rivera, C. Lindsay, J.F. Codocedo, I. Morel, C. Pinto, P. Cisternas, F. Bozinovic, N.C. Inestrosa, Andrographolide recovers cognitive impairment in a natural model of Alzheimer's disease (Octodon degus), Neurobiol. Aging. 46 (2016) 204–220. doi:10.1016/j.neurobiolaging.2016.06.021.

[249] J. Geng, W. Liu, Y. Xiong, H. Ding, C. Jiang, X. Yang, X. Li, A. Elgehama, Y. Sun, Q. Xu, W. Guo, J. Gao, Andrographolide sulfonate improves Alzheimer-associated phenotypes and mitochondrial dysfunction in APP/PS1 transgenic mice, Biomed. Pharmacother. 97 (2018) 1032–1039. doi:10.1016/j.biopha.2017.11.039.

[250] X. Cai, Z. Fang, J. Dou, A. Yu, G. Zhai, Bioavailability of Quercetin: Problems and Promises, Curr. Med. Chem. 20 (2013) 2572–2582. doi:10.2174/09298673113209990120.

[251] W. Wu, Y. Wang, L. Que, Enhanced bioavailability of silymarin by self-microemulsifying drug delivery system, Eur. J. Pharm. Biopharm. 63 (2006) 288–294.

doi:10.1016/j.ejpb.2005.12.005.

[252] A. Das, Y.M. Gangarde, V. Tomar, O. Shinde, T. Upadhyay, S. Alam, S. Ghosh, V. Chaudhary, I. Saraogi, Small-Molecule Inhibitor Prevents Insulin Fibrillogenesis and Preserves Activity, Mol. Pharm. 17 (2020) 1827–1834. doi:10.1021/acs.molpharmaceut.9b01080.

[253] A. Shamsi, A. Ahmed, M.S. Khan, F.M. Husain, B. Bano, Rosmarinic acid restrains protein glycation and aggregation in human serum albumin: Multi spectroscopic and microscopic insight - Possible Therapeutics Targeting Diseases, Int. J. Biol. Macromol. 161 (2020) 187–193. doi:10.1016/j.ijbiomac.2020.06.048.

[254] P. Patel, K. Parmar, M. Das, Inhibition of insulin amyloid fibrillation by Morin hydrate, Int. J. Biol. Macromol. 108 (2018) 225–239. doi:10.1016/j.ijbiomac.2017.11.168.

[255] H. Eskandari, M. Ghanadian, C. Noleto-Dias, C. Lomax, A. Tawfike, G. Christiansen, D.S. Sutherland, J.L. Ward, H. Mohammad-Beigi, D.E. Otzen, Inhibitors of α-Synuclein fibrillation and oligomer toxicity in Rosa damascena: the all-pervading powers of flavonoids and phenolic glycosides, ACS Chem. Neurosci. (2020). doi:10.1021/acschemneuro.0c00528.

CHAPTER 2

Background of Work and Objectives

2. Background of Work and Objectives

Protein fibrillation and amyloid fibrils deposition in tissue sites are some of the activators in degenerative diseases. Protein fibrillation is triggered by various factors including, stress such as heat and oxidative conditions, cellular mutations, aging and other translational complications [1,2]. Proteins fibrillations originate with misfoldings due to molecular centres and are precipitated by various regulators. Thermodynamically unfavourable intermediates lead to disordered aggregates, changing slowly into toxic amyloid fibrils. Guiding forces in molecular levels are non-covalent interactions including electrostatic interactions, H-bonding, aromatic π interactions and van der Waals forces. Plasma proteins that form amyloid fibrils share a common feature of high β sheet conformation, representing a clue of the proteins secondary structural requirements for fibril formations [3]. Typical plasma proteins are often used to investigate into amyloid fibril formations *in vitro*. This leads to better understanding of the initiation, propagation, molecular interactions and/or inhibition and possible reversal of protein fibrillation. Protein fibrillation can proceed in two distinct mechanisms: misfolding of monomeric intermediates, leading to seedlings of fibrils, which exhibiting lag phase in the protein fibrillation kinetics; and misfolding of oligomeric intermediates appearing as seeds with absence of lag phase in the fibrillation kinetics curve [1]. In biological conditions, amyloid fibrils appear over a period of time. Flexible protein structures are often incited into nucleation for stress conditions and regulators leading to in-digestible fibril formations.

Amyloid fibrils originate from normal soluble proteins and associate in diseases that are termed together as amyloidosis. The 'amyloid' term was but coined way back in 1854 in order to describe, iodine stained waxy deposits seen in liver autopsy of patients suffering from chronic inflammatory diseases. Rudolf Virchow (1854) considered these as starchy (Latin: *amylum)* carbohydrates [3,4]. However, these were soon turned out to be nitrogen rich β structured protein fibrils resistant to proteolysis. Therapeutic leads for intervening in amyloid fibrillation are presently immerging in Alzheimer's disease and Parkinsonism. Depicting amyloid fibrillation *in vitro* and effective interruption are some of the fundamental challenges for mitigating a range of dilapidating disease conditions.

This work intends to address a few of the challenges in development of agents capable of interceding protein misfolding for subsequent extenuation/ inhibition of fibrillation. Specific compounds isolable from plants were in prime focus rather than the synthetic small molecules and chaperones. This is due to their structural diversity, druggability and low toxicity. Site specific interactions occurring between the investigating molecules and human serum albumin (HSA) were explored and their effects in protein fibrillation interfaces were determined. Plant bioactives were also tethered on gold nanosurface and their effects on protein fibrillation were studied. Surface engineered biopolymer nanoparticles were experimented in plasma protein interface to perceive their role in fibrillation and corona formation. Knowledge of the nanosurface chemistry were applied for development of new generation nanomedicines. The plant compound with potent fibrillation inhibitory property was selected as the bioactive payload for the engineered nano-carriers. Therapeutic efficacy of the nanomedicines developed through a synchronized design was further evaluated in selective hepatic tissue degenerative conditions.

The work was planned to progress in different well connected phases. The studies embodied in the dissertation can be consolidated under four sub-headings.

1. Representative compounds of plant origin, andrographolide, silybin and quercetin were explored individually in order to elucidate their effects on HSA fibrillation *in vitro*. Molecular interactions between the compounds and the model protein were studied in depth. Effects of these compounds on protein fibrillation were studied through a combination of spectroscopy and microscopy techniques.

2. Bioactive molecule tethering techniques on gold nanoparticles were applied. Low temperature reactions were used for *in situ* synthesis of gold nanoparticles. Different plant bioactives were explored for minimized toxicity and understanding efficacy in nanoscale formations. Effects of differently tethered gold nanoparticles on protein fibrillation were carefully studied.

3. Understanding the behaviour of nanoparticles in physiological conditions appeared essential prior to, designing nanomedicines for translational applications. A study on the effects of nanosurface chemistry in protein misfolding and corona formation was thus undertaken. PLGA copolymer was used as a base for nanoparticles preparation. Interactions between differently functionalized nanoparticles and target plasma protein were observed and the role of surface-stabilization on corona effects were investigated.

4. Nanoparticles stabilized using Pluronics was applied for designing of biopolymer nanoparticles capable of slow delivery of bioactive payload. The plant bioactive with maximum fibrillation inhibitory capacity was subsequently used. Biochemical effects of new generation nanomedicines were explored against paracetamol overdose hepatotoxic conditions.

The work mostly took an effect on understanding potential bioactives efficacy on a major plasma protein fibrillation under experimental conditions. Studies on plant bioactives for fibrillation inhibition and preformed fibrils disruption were undertaken. Bioactive cargo loaded nanoparticles were further investigated against hepatotoxic conditions in animal models. Effects of functionalized gold nanoparticles and surface modified biopolymer nanoparticles on protein fibrillation and bimolecular corona formations were also studied in depth.

CHAPTER 3

Effects of Andrographolide on Protein Fibrillation

3.1. Introduction

Andrographolide (AG) is a labdane diterpenoid extracted from the *Andrographis paniculata* leaves. Extracts of this plant, leaves and roots were used for long in India, China, Malaysia, Thailand and other Southeast Asian countries in order to mitigate various ailments like malaria, diabetes, viral infections, hypertension and cancer [5–8]. AG consists of an exocylic γ-lactone moiety and is an electrophilic plant bioactive that can secure both covalent and non-covalent bonds with targets molecules [9–11]. AG interacts with several inter- and intracellular components as a bipolar compound, therefore triggering many biological responses. AG exhibited efficacy in amyloidogenic disease conditions such as Aβ neurotoxicity, rheumatoid arthritis, diabetes and related complications [12–14]. Comprehensive molecular interactions with amyloidogenic proteins which underlie its therapeutic effect are however very unclear. It was therefore envisaged that studying the effects of andrographolide on protein fibrillation would help recognize the molecular interactions behind its pharmacological properties.

This part of the work was organized to illustrate the effect of different concentrations of andrographolide on protein fibrillation. The toxicological profiles of different concentrations were primarily assessed through cell viability studies. The process of HSA fibrillation has been monitored using a combination of various biophysical techniques. Further, efforts were directed to elucidate the interactions between andrographolide and HSA protein using spectroscopic experiments and *in silico* studies.

3.2. Experimental

3.2.1. Protein solution and fibril formation

HSA stock solution (100 µM) was constituted by dissolving lyophilized HSA (Sigma Aldrich, US) in phosphate buffer solution (10 mM PBS, pH 7.4). The solution was filtered using a Millipore filter (Millipore, USA) having pore size of 0.44 µm before use. Concentration was ascertained spectrophotometrically, using molar extinction coefficient 35219 M^{-1} cm^{-1} at 280 nm. HSA solutions were then incubated at 65±2 ^{0}C for induced amyloid fibril formation alone or with different ratios of AG (Natural Remedies, India) (HSA:AG:: 1:0, 1:0.5, 1:1 and 1:2) in an orbital shaker bath set at 80 rpm.

3.2.2. Dynamic light scattering studies

The HSA fibrillating samples (3.2.1) were diluted to 2 μM in particle-free water and the average size was obtained using a dynamic light scattering (DLS) facility, zetasizer Nano ZS (Malvern Instruments, UK) set at 25±2 °C. The hydrodynamic radii (R_h) of the aggregates were calculated in from an in-built intensity autocorrelation function based on translational diffusion co-efficient, and was presented (figure 3.1) with respect to time.

3.2.3. Thioflavin T fluorescence spectroscopy

The fibril formation process was further studied through Thioflavin T (ThT) fluorescence spectroscopy. The ThT stock solution (200 μM) was prepared in 10 ml of phosphate buffer containing 0.1 M NaCl and 0.02 % NaN_3 [15]. HSA samples with or without AG were allowed to incubate with freshly prepared ThT dye solutions at room temperature for 20 minutes. The protein: dye ratio was fixed at 5:1 [16]. Fluorescence intensities were recorded (figure 3.2) in a Perkin-Elmer LS-5 spectrofluorimeter (PerkinElmer Inc., US), having exciting and emission wavelengths set at 450 and 480 nm. All blanks, HSA protein and AG bioactive were checked for no-responsiveness in ThT fluorescence domain.

3.2.4. Cytotoxicity evaluations

In vitro cytotoxicity of AG was evaluated through trypan blue exclusion assay [17]. Briefly, HaCaT cells were cultured at a density of 1×10^5 per well in 24-well plates for 24 hours. The cells were either untreated or treated with different doses of AG (5-50 μM) for 24 hours. The cells were then harvested and trypan blue solution (0.4 %) was added to the cell suspensions in a ratio of 1:1. Finally the number of total and dead cells were counted using a hemocytometer. Data were averaged from a set of three replicates and the percentages of viable cells were calculated.

3.2.5. HSA Andrographolide interaction studies

UV absorption spectra of HSA (5 μM) incubated with varying concentrations of andrographolide (AG, 0-4 μM) were recorded in UV-Visible spectrophotometer UV-1800 (Shimadzu Corporation, Japan) using matched quartz cuvettes of 1 cm path length. Solutions were first equilibrated at 37±2 °C for 30 minutes and the spectra in a range 250 nm to 350 nm were then collected for studies.

Fluorescence emissions in a range of 300 nm to 500 nm were also recorded in a spectrofluorimeter, set with an excitation wavelength of 280 nm. The instrument was fitted with a water bath circulator controlled by Neslab RTE 100 thermostat. Fixed concentration of HSA protein (5 µM) was titrated with different concentrations of AG and the fluorescence quenching spectra obtained at 37±2 °C. No emission due to AG was also recorded upon excitation at 280 nm.

3.2.6. Molecular modelling studies

The structure of AG bioactive was developed *in silico* [18]. Induced fit docking (IFD) protocol (Schrödinger, LLC, New York, NY. 2013) was run in three consecutive steps. The bioactive was first docked into a rigid receptor model with scaled-down van der Waals (vdW) radii. A vdW scaling of 0.5 was used for both the protein and bioactive nonpolar atoms. A constrained energy minimization was carried out on the HSA structure (Protein Data Bank id: 2BXP, accessed on December, 2016) keeping it closer to the crystal structure and removing steric contact hindrance. Energy minimization was performed using Optimized Potentials for Liquid Stimulation (OPLS2005) force field [19] with implicit solvation model until the default criteria were satisfied. The centroid of the co-crystal ligand was used for defining the location of the binding site and the dimension of energy grids for the initial docking was fixed to 25 Å. A Grid-based Ligand Docking with Energetics mode (Glide XP, Schrödinger, LLC, New York, NY) was applied for the primary docking, and 20 ligand poses were retained for structural refinements [20]. In the next step, Schrödinger's Prime module was used to generate induced-fit HSA–AG complexes. Each structure from the previous step was subjected to side-chain and backbone refinements. All residues with at least one atom located within 5.0 Å of each corresponding ligand pose were included in the Prime refinement. The refined complexes were then ranked by Prime energy, and receptor structures within 30 kcal/mol of minimum energy structure approached for the final round of Glide docking. In the last step, each ligand was re-docked into every refined low-energy receptor structure generated in the second step using the Glide XP. Induced fit docking was completed using the IFD docking module of Schrödinger software. Best dock pose of AG was determined based on the interactions shown as well as Glide XP and IFD scores.

3.2.7. Transmission electron microscopy (TEM)

Microscopic investigations were performed to analyse the morphological attributes of the HSA fibrils formed in presence and absence of AG. 10 µl of the sample was placed on a carbon coated copper grid (Ted Pella Inc., US) and was negatively stained with 2 % w/v uranyl acetate solution. Electron micrographs were obtained from a high resolution JEOL JEM 2100 electron microscope (JEOL, Japan) operated at 120 kV.

3.2.8. Circular dichroism (CD)

Far UV-CD spectra were recorded between wavelength 190 to 250 nm on a JASCO J-815 CD spectropolarimeter (JASCO Corporation, Japan) set with a scan speed of 100 nm/min and bandwidth of 1 nm. Samples (section 3.2.1) were placed in a 1 mm quartz cell and all measurements were recorded at 25±2 °C (figure 3.6). Each spectrum was baseline corrected and ellipticity was obtained from an average of three scans acquired.

3.2.9. Statistical analysis

All experiments were conducted in triplicates and data have been presented as mean ± standard deviations (S.D.). Student's t-test was done for comparative analyses of means. Differences were considered significant only if $P<0.05$.

3.3. Results and discussion

3.3.1. Effect of AG on HSA fibril formation

HSA was co-incubated with or without plant bioactive AG at 65±2 °C and the hydrodynamic radii of HSA aggregates during the incubation phases were recorded using DLS (figure 3.1). The samples were diluted to diminish protein interactions and viscosity effects [21]. R_h of untreated HSA was 4.1±0.3 nm, and that was found similar to earlier reports [22]. The same value escalated up to 218±4.2 nm upon incubation at 65±2 °C, signifying the partial unfolding of serum albumin tertiary structures [23] and formation of higher molecular weight aggregates. The growth of protein aggregates when treated with bioactive AG in ratios of 1:0.5, 1:1, and 1:2 were however limited to 166±3.7 nm, 123±3.1 nm and 85±2.8 nm, in 4 hours of incubation. The data from DLS experiments expressed concentration dependent changes in the hydrodynamic radii of HSA protein upon incubation with AG. This was likely due to protein unfolding and conformational shifts during oxidation/ reduction, or AG-HSA complex formations [24].

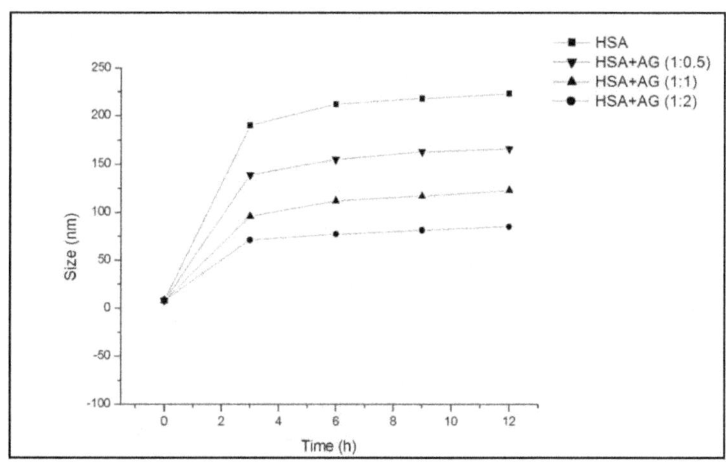

Figure 3.1. HSA fibril formation over time, in absence or presence of different ratios of AG as measured by DLS at 25±2 °C.

Thioflavin T probe binds specifically to amyloid fibrils and not to the precursor proteins or aggregates [25]. ThT fluoresces pronouncedly near 480 nm, when bound to the hydrophobic cavity of HSA [26]. ThT response from the untreated HSA increased steadily after a lag period of 1 hour and reached saturation at 6 hours (figure 3.2). This was in agreement with the polymerization models for amyloidogenic proteins [27]. A concentration dependent inhibitory effect was observed when HSA was incubated with AG in different stoichiometric ratios. A significant decrease of ThT intensity at saturation phase was noted at higher ratios of 1:2 ::HSA:AG as compared to a sub-stoichiometric ratio of 1:0.5 for HSA:AG. The bioactive was able to repress fibril formation by ~64 % at 1:2 of HSA:AG ($P<0.05$).

The data were also fitted in sigmoidal equation model (equation 3.1) in order to understand the lag time information (Microcal Origin 6.0).

$$F = F_{min} + \frac{F_{max}}{1+e^{-[\frac{t-t_0}{\tau}]}} \tag{3.1}$$

Where, Y, the ThT fluorescence response, F_{min}, minimal ThT fluorescence response, F_{max}, maximum ThT fluorescence response, t, incubation time and t_0, was the time to achieve 50 % of maximal fluorescence. The apparent first order rate constant for HSA fibril growth was expressed as $1/\tau$, while the lag time was calculated as $t_0-2\tau$.

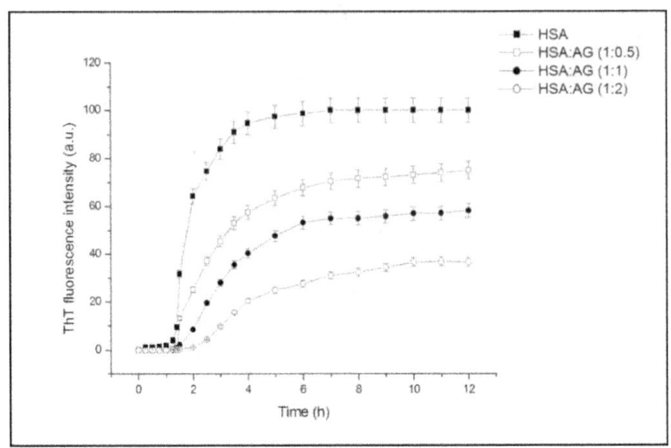

Figure 3.2. ThT growth curve of HSA in absence and presence of different ratios of AG. ThT intensities were expressed as mean±S.D. (n=3).

Detailed analysis (table 3.1) revealed that AG increased the lag time for HSA fibrillation and repressed the protein polymerization rate in a concentration dependent manner. This showed that AG caused the delay of HSA monomers aggregation and primary nucleation.

Table 3.1: Lag time and apparent rate constant of HSA fibrillation

HSA:AG	Lag time (hour)	Apparent rate constant (hour^{-1})
1:0	0.68 ±0.066	2.68
1:0.5	0.88 ±0.132	1.24
1:1	1.37 ±0.081	1.21
1:2	1.44 ±0.158	0.86

Results were expressed as mean±S.D. (n=3).

3.3.2. Cell viability studies

Cell viability studies showed that AG did not cause significant death of HaCaT cells at 5.0 – 50.0 µM concentration ranges (figure 3.3). This observation was found in line with the earlier works and confirmed that AG did not exert any significant effect to the normal cells [28,29].

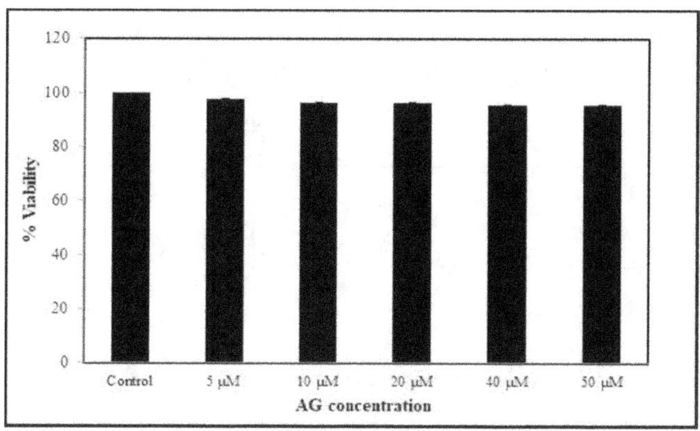

Figure 3.3. Cell viability assay of different concentrations of AG on HaCaT cell lines. Results were expressed as mean±S.D. (n=3).

3.3.3. Characterization of binding interaction between HSA and AG

UV absorption spectroscopy is an effective tool to perceive the serum albumin structural changes and specific interactions with different molecular species [30]. The absorption spectrum of HSA (figure 3.4A (b)) features a distinctive peak at 278 nm owing to amino acid residues such as phenylalanine, tryptophan and tyrosine. AG was lacking any significant absorbance in that ranges. UV absorbance intensities were recorded with increasing concentrations of AG (c-f). This indicated non-covalent interactions due to molecular diffusion interplayed between HSA and AG. Such interactions were likely due to hydrophobic attachments of AG with phenyl ring of aromatic amino acids present in HSA binding cavities. These observations primarily suggested changes in the protein microenvironment during ground-state complex formation [31,32]. The binding constant value (K) was calculated (equation 3.2) as described by Benesi and Hildebrand [33]:

$$\frac{1}{A_{obs}-A_o} = \frac{1}{A_c-A_o} + \frac{1}{K(A_c-A_o)[AG]} \tag{3.2}$$

Where, A_{obs} was the observed absorbance of HSA solution in presence of different concentrations of AG at 278 nm. A_o and A_c were the absorbance values for HSA and HSA-AG complexes. Concentration of AG expressed in mol/l was used as [AG]. The K value was derived from a graphical plot (figure 3.4A inset) and was found to be 4.76 x 10^4 M^{-1}. The free energy of interaction was also derived from the binding constant (K) (equation 3.3).

$$\Delta G = -RTlnK \tag{3.3}$$

Where ΔG, the binding free energy, R, gas constant 1.987 cal/mol and T was 310.15 K throughout. The value of ΔG was found to be -6.6 kcal/mol. The negative value of ΔG confirmed the spontaneity of bioactive AG in HSA binding.

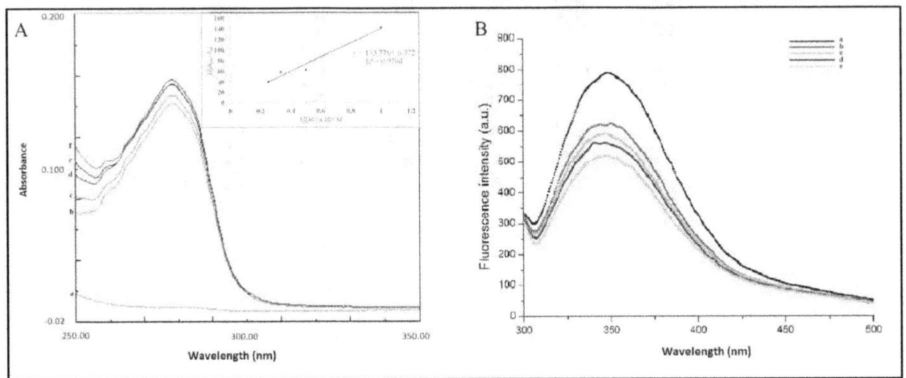

Figure 3.4.(A) Absorption spectra of AG (a), HSA in absence (b) and presence of different concentrations (1-4 μM) of AG (c–f). Inset: Benesi and Hildebrand plot; (B) fluorescence spectra of HSA in presence of different concentrations (0-4 μM) of AG (a-e).

Intrinsic fluorescence property of HSA is specific due to the presence of tryptophan residue (TRP 214) in the IIA subdomain (Sudlow's site I) [16]. This was often attenuated during interaction with a small molecule in its microenvironment [34]. HSA demonstrated a characteristic emission peak at 350 nm, upon being excited at 280 nm. A steady decrease of intensity of the emission band of HSA at 350 nm was observed (figure 3.4B) with increasing concentrations of AG. This was due to the inclusion of AG into the binding site of HSA and subsequent formation of HSA-AG complex. Fluorescence studies have helped further to ascertain HSA hydrophobic site specific interactions for AG. No change of maximum emission wavelength suggested that the native protein conformation possibly remained intact, upon binding with AG [35].

3.3.4. Molecular docking and stimulation

The best dock pose (figure 3.5B) presented the Glide XP and IFD score as -11.95 and -1299.82 kcal/mol. Glide XP score indicated binding affinity of ligand towards the protein target. IFD score was a combination of Glide XP score and the prime refinement energy of binding cavity. AG formed hydrogen bond interactions with LYS 195, ALA 210, SER 454 and VAL 482 of the protein (figure 3.5B) due to diterpenoid α-alkylidene γ-butyrolactone scaffold and the hydroxyl group at C 14. Residues LYS 195 and SER 454 acted as hydrogen bond acceptors for hydroxymethyl and hydroxyl groups occurring at a distance of 2.95 Å and 2.31Å. Residue ALA 210 established hydrogen bonding at a distance of 1.77 Å with the hydroxyl group, while, the carbonyl group formed a hydrogen bond 2.33 Å with the protein VAL 482 residue. The decahydronaphthalene ring of AG also interacted with TRP 214 through hydrophobic interactions. The *in silico* studies corroborated well with the fluorescence results (3.3.3) and also presented a good evidence explaining the quenching of intrinsic fluorescence responses at 350 nm. The site-specific interactions thus mediated the stabilization of the native protein structure and that validated the results obtained from UV absorption and fluorescence spectroscopy. Molecular docking results corroborated well with the *in vitro* experiments. Apparently, AG attached to the albumin binding cavity and concentration dependently affected the hydrogen bonding networks. This was expected to cause minor orientation changes without affecting much in-protein conformation or helicity.

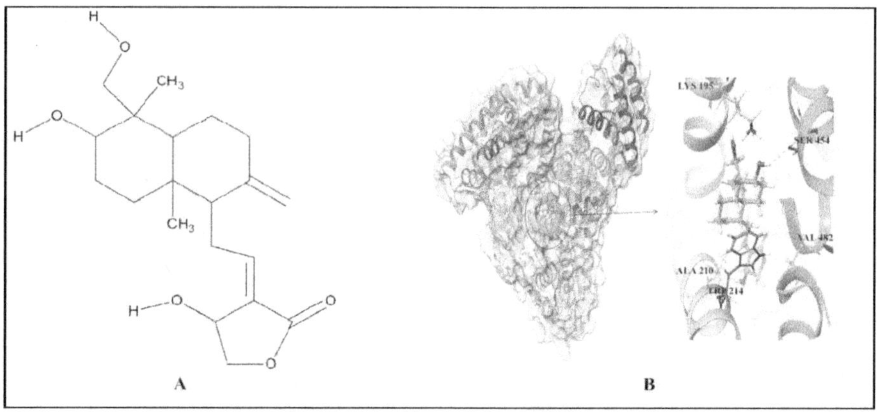

Figure 3.5. (A) Chemical structure of AG (B) Surface view of the crystal structure of HSA bound to the docked AG showing H-bond (yellow dotted lines) interactions with binding site residue.

3.3.5. *Effect of AG on HSA secondary structures*

Far UV CD spectroscopy is a sensitive tool to estimate the progressive loss of protein structural integrity and β sheets formations during fibrillation [36]. Native HSA samples expressed 73 % α helix with a characteristic CD spectrum giving rise to two negative bands at 208 nm and 222 nm (figure 3.6A). The negative peaks at 208 and 222 nm arise from the $\pi \rightarrow \pi^*$ and $n \rightarrow \pi^*$ transitions of carbonyl groups present in the polypeptide chains [37]. HSA fibrillation (3.3.1) caused loss of negative ellipticity at 208 and 222 nm through disruption of α helix structures and associated β sheets formations (44.5 %). AG concentration dependently restricted the evolution of β sheets. An online server BeStSel (ww.bestsel.elte.hu) was used for secondary structures estimations [38]. The β sheet contents relative percentage were 34.6 % when HSA was incubated with AG at 1:1, while that value lowered to 27.4 % for 1:2 (figure 3.6B). The CD results were found to be in agreement with the spectroscopic interpretations. AG was considerably effective in retarding HSA fibrillation through stabilization of native protein structure.

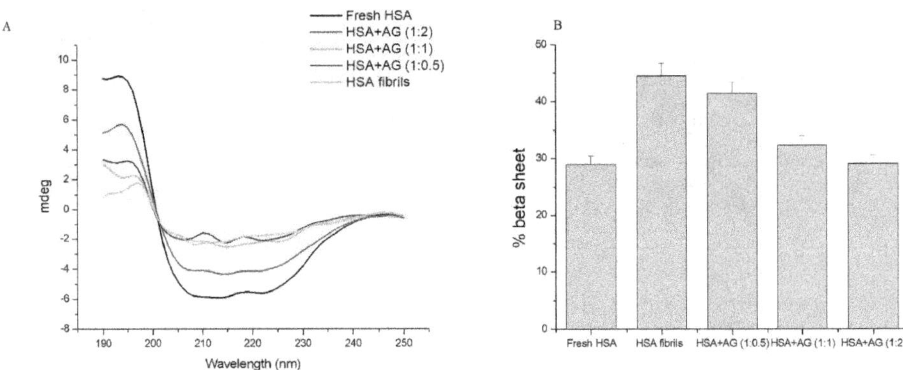

Figure 3.6. (A) CD spectra and (B) histograms of % β sheet content in HSA solution in absence and presence of AG.

3.3.6. Effect of AG on HSA fibril morphology

Morphological examinations showed that in absence of AG, HSA developed into long, un-branched and compact structures upon incubation at 65±2 °C (figure 3.7A). The width of matured fibrils was found to be 30 nm. In contrast, HSA samples incubated with higher ratio of AG (1:2) gave rise to amorphous aggregates rather than well-defined fibrils (fig. 3.7D). HSA solutions containing lower ratios of AG produced indistinct fibrillar structures (fig. 3.7B, C), indicating that AG suppressed maturation of amyloid fibrils through stabilization of the protein structure at pre-fibrillar phases [39]. This was probably due to inclusion of AG into HSA subdomain binding site, leading to altercations in hydrogen bonding networks. The binding sites for AG in the protein appeared specific which led to minor unfolding and caused hindrance to nucleation during protein fibrillation. The TEM observations confirmed the repressive effect of AG on HSA fibrillation and as a function of concentration.

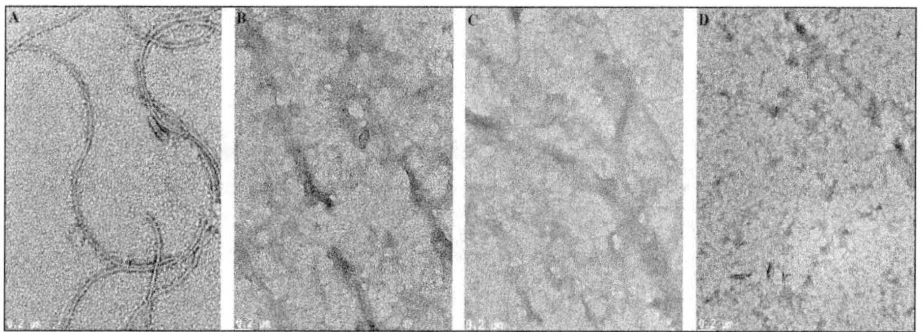

Figure 3.7. TEM studies of 2 µM HSA solution incubated alone (A) and with AG at ratios of (B) 1:0.5, (C) 1:1, (D) 1:2. Scale: 0.2 µm.

3.4. Summary

AG suppressed human serum albumin fibrillation in a concentration-dependent manner. A maximum of ~64 % inhibition was recorded at 1:2 ratio of HSA:AG. Findings also indicated that AG stabilizes native protein α helical structure and restricts β sheet formations. Non-covalent interactions between AG and HSA were recorded though multi-spectroscopy and *in silico* experiments. This work thus demonstrated that plant diterpenoid andrographolide concentration dependently inhibited HSA fibril formation through site specific interactions.

CHAPTER 4

Effects of Quercetin on Protein Fibrillation and Preformed Fibrils

4.1. Introduction

Quercetin (QR) is a dietary flavonoid ubiquitously distributed in vegetables, fruits, red wine and tea. QR, is one of the most potent antioxidants of plant origin [40]. QR (3,5,7,3',4', pentahydroxy flavone) has exhibited several biological effects such as anti-inflammatory, anti-allergy, anti-fibrotic, antiviral and cardioprotective activities [41–43]. The flavonoid is also one of the best known phytoestrogens capable of reducing risks to various cancer conditions [44]. Biochemical studies indicated that QR exerted neuroprotective effects against chemical induced toxicity, enhanced reduced glutathione levels and increased the resistance of the cells to oxidative stress [45]. QR is known to improve cognitive and emotional functions in Alzheimer's disease through reduction of β amyloid and tau protein levels in the brain tissues [46]. Neuroprotective effects of QR are also apparent in Parkinson's disease models [47]. QR in addition expressed favourable effects in diabetes and related complications, including obesity and vasoconstriction [48,49].

QR has been intently studied for its activity on cellular redox status. The free radical scavenging effects correlated well with the protective effects of QR in oxidative stress induced amyloid protein fibrillations [50,51]. In another observation, QR repressed insulin fibrillar assembly formations and may have helped in injection-localized amyloidosis [52]. QR exerts very unique and protein site specific interactions. Recently, QR has proved as a very successful agent against influenza virus protein and COVID-19 [43,53]. Understanding QR interactions with a prime biological target protein was therefore considered for unravelling its effects on disease modifications and molecular pathophysiology.

This part of the work was intended to explore the effects of QR on HSA fibrils. Molecular level interactions of QR with HSA were studied both in wet lab and *in silico* modelling experiments. QR biopharmaceutical uptake is low and seemingly HSA is one principal carrier in the systemic circulations. The implications of molecular interactions on progression of protein fibrillation appeared revelling. Effects of QR on matured HSA fibrillar assemblies were further studied in this case. Observations were interesting. A combination of computational studies, biophysical laboratory techniques and electron microscopy studies have helped to arrive at some of the concluding observations on flavonoid structure specific effects on protein fibrillations.

4.2. Experimental

4.2.1. Formation and disruption of HSA fibrils

20 µM HSA solutions (section 3.2.1) were incubated alone or in presence of varying concentrations of QR, (HSA:QR ratio:: 1:0 to 1:10) at 65 ± 2 °C in an orbital shaker bath set at 80 rpm. The extent of fibril formation was tested through various biophysical techniques.

Fibril disruption experiments were carried out in addition. For that, HSA solutions were diluted to 2 µM with PBS. HSA fibrils were allowed to mature fully over a period of 120 hours at 65 ± 2 °C [54]. The matured fibrils were then allowed to incubate with different concentrations of QR (0 to 30 µM) at 37 ± 2 °C for 48 hours without any agitation and occurrence of disintegration, if any, was monitored using biophysical techniques.

4.2.2. Dynamic light scattering studies

HSA samples with or without QR were diluted with particle-free water so that the final concentration of protein achieved in the solutions was 2 µM. The average hydrodynamic radii were then studied using a zetasizer (section 3.2.2) at 25 ± 2 °C.

4.2.3. Thioflavin T fluorescence spectroscopy

HSA samples with or without QR were allowed to incubate with ThT dye solution for 20 minutes in the dark (section 3.2.3) [16]. Protein: dye ratio was maintained at 1:5. The fluorescence intensities were then recorded in a spectrofluorimeter, having excitation and emission wavelengths fixed at 450 and 480 nm. Fluorescence response for QR was also recorded and non-responsiveness in ThT fluorescence region was confirmed.

4.2.4. Cytotoxicity studies

Cytotoxicity of QR was evaluated through trypan blue exclusion assay following cell culture protocols as described earlier (section 3.2.4). Cultured HaCaT cells were either treated or left untreated with different doses of QR (5-50 µM) for 24 hours. The cells were then harvested and 0.4 % trypan blue solution was added in a ratio of 1:1. Finally the numbers of live and dead cells were counted and the averaged of three replicate sets were used for evaluations.

4.2.5. HSA Quercetin interaction studies

Quercetin (QR, 15.1 mg, Sigma Aldrich, USA) was dissolved in a minimum amount of DMSO and the volume adjusted to 10 ml with phosphate buffer culminating in a 5 mM stock solution. UV absorption spectra of HSA (5 μM) incubated with different concentrations (0-4 μM) of QR were then recorded in UV-Visible spectrophotometer and the methodology was described earlier in section 3.2.5. Solutions were equilibrated at 37±2 °C for 30 minutes before spectroscopy observations.

Fluorescence emission spectra in range of 300 nm to 500 nm were also recorded using a spectrofluorimeter, and an excitation wavelength of 280 nm was used. Width of both excitation and emission slits were fixed at 5 nm. Protein solution (5 μM) was incubated with bioactive QR (0-4 μM) and the spectra were studied after incubation at 37±2 °C. QR alone did not fluoresce upon excitation at 280 nm.

4.2.6. Molecular modelling studies

In silico studies were performed in order to augment experimental observation and for further understandings. The molecular structure of QR was developed using LigPrep tools (Schrödinger, LLC, New York, NY. 2013) and energy was optimized using OPLS2005. IFD protocol was then implemented using the Prime module following steps as described in section 3.2.6. The IFD and Glide XP scores for the best pose were selected for subsequent analysis. The molecular interactions were also visualized using 2D interaction diagram utility tools and Schrödinger suite in Maestro user interface (Schrödinger, LLC, New York, NY. 2013).

4.2.7. Transmission electron microscopy

Morphology analysis of HSA fibrils formed in presence or absence of QR were carried out in an electron microscope (section 3.2.7). Samples placed on a carbon coated copper grid were negatively stained with uranyl acetate and micrographs were obtained.

4.2.8. Statistical analysis

All experiments were conducted in triplicates and data presented as mean±S.D. The mean values were analysed through student's t-test and differences were considered significant when $P<0.05$.

4.3. Results and discussion

4.3.1. Effect of QR on HSA fibril formation

Evidences on the effects of different stoichiometry ratios of QR on HSA fibrillation were initially collected from DLS experiments. DLS was also applied to obtain a general idea about the state of the protein (native or aggregated) under consideration [55]. HSA solutions incubated alone at 65±2 °C for 12 hours exhibited an increase in hydrodynamic radii over time (figure 4.1). This was likely due to induced irreversible aggregation. The R_h values of the protein aggregates treated with QR in ratios of 1:0.5, 1:1, 1:2 and 1:3 were but progressively limited to 147±11, 124±8, 106±5 and 85±2 nm, respectively. Observations suggested that the changes in the hydrodynamic radii of the protein aggregates were likely depended upon the concentration of QR in solution. Furthermore, the QR concentration dependent size effects were very parallel from 6 hours of incubation, indicating a subtle equilibrium near the time point. No major change in the aggregate size was however observed when the ratio exceeded 1:3. Additionally, it was observed that presence of QR in solution affected the growth phase of HSA fibrils, and the QR co-incubated fibrils might not have reached their maximum size by 6 hours of incubation.

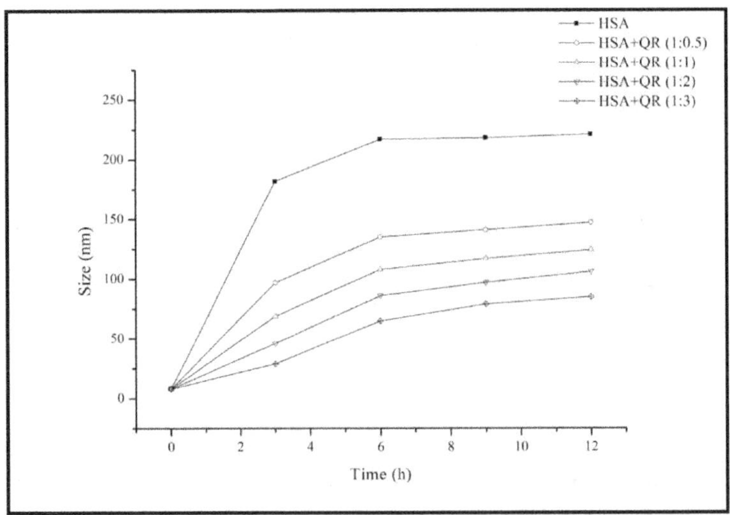

Figure 4.1. Hydrodynamic radii of HSA incubated with or without different ratios of QR at 65±2 °C.

Standard thioflavin T spectroscopy studies were performed further in order to obtain quantitative results of the inhibitory effects of QR on HSA fibrillations. An increase in the ThT fluorescence emission response upon incubation of HSA solution at high temperature relates to formation of protein fibrils [56]. Samples containing bioactive QR exhibited a steady decrease of ThT intensity with increasing concentrations of QR, thus confirming the fibril inhibitory of the molecule (figure 4.2). Designated proteins aggregate under stress conditions in form of highly structured amyloid fibrils. The ThT dye selectively binds to the β sheets structures and presents a considerable increase in fluorescence intensity and a red-shift in the excitation-emission maxima. The ThT binding to the amyloidogenic fibrils was explained in 'Krebs model' [57]. The dye is known to insert itself in the β-sheets grooves formed by the amino acids side chains due to affinity. This results in a rotation of benzothiazole and aminobenzoyl rings in the excited state causing enhanced fluorescence. The HSA:QR ratio for the maximum effective inhibition of fibrillation appeared to be 1:3, and that was able to reduce the β sheets fibrillar content by 44 %. A combination of DLS and ThT fluorescence experiments thus demonstrated concentration-dependant HSA fibril inhibitory capacity of QR under thermally induced denaturing conditions in phosphate buffers.

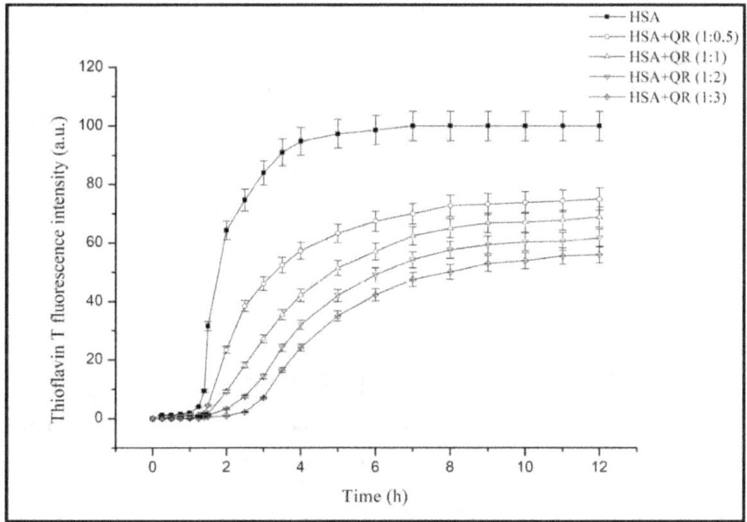

Figure 4.2. ThT growth curve of HSA in absence and presence of different ratios of QR. ThT intensities were expressed as mean±S.D. (n=3).

Once the maximum effective inhibitory ratio was estimated, the effect of QR at that level (6 µM) was further studied on the HSA fibrillation kinetics. A significant deviation from the sigmoidal curve of HSA fibrillation standard was observed when the protein was incubated with QR in the beginning. The data obtained from the kinetics studies were fitted further in the sigmoidal equation (3.1) in order to derive the lag time information. The progression of HSA fibrillation was consistent with the nucleation dependant polymerization model of amyloidogenic proteins featuring in an initial lag phase, short elongation phase, and a saturation phase [58]. At a concentration 6 μM, QR extensively increased the lag time of HSA fibrillation from 0.68±0.07 hour for HSA alone to 1.94±0.19 hours (P<0.05). The rate of fibrillation of HSA was concurrently lowered from 2.68 hour^{-1} to 0.6 hour^{-1} in presence of QR. These findings indicated that the bioactive QR interacted with the protein molecules, delayed nucleation and oligomers formations and extended the lag phase.

$$F = F_{min} + \frac{F_{max}}{1+e^{-[\frac{t-t_0}{\tau}]}} \tag{3.1}$$

4.3.2. Effects of QR on HSA fibril morphology

A detailed electron microscopic analysis was intended to study the morphology of the fibrils and related aggregates formed during the incubation of HSA at 65±2 °C (figure 4.3). TEM micrograph of HSA incubated alone for 1 hour (A1) revealed presence of small globular entities. Micrographs of samples from different time intervals (A2 and A3) showed generation of protofibrils and their subsequent maturation to well-defined fibrils (A4). Micrographs of HSA incubated with QR (6 μM) at similar conditions exhibited indistinct aggregates of albumin sampled just after 4 hours (B2). Dark, elongated patches were prominent after 6 hours of incubation (B3), along with existence of aggregates without fibrillation. After 12 hours of studies, the micrographs (B4) consistently appeared amorphous and morphologically dissimilar from that in A4. The formation of oligomers, protofibrils and fibrils was well associated with amyloid deposition in physiological system [59]. Absence of similar defined structures in the electron micrographs corroborated with other instrumental observations and suggested a profound inhibitory effect due of the bioactive flavonoid QR.

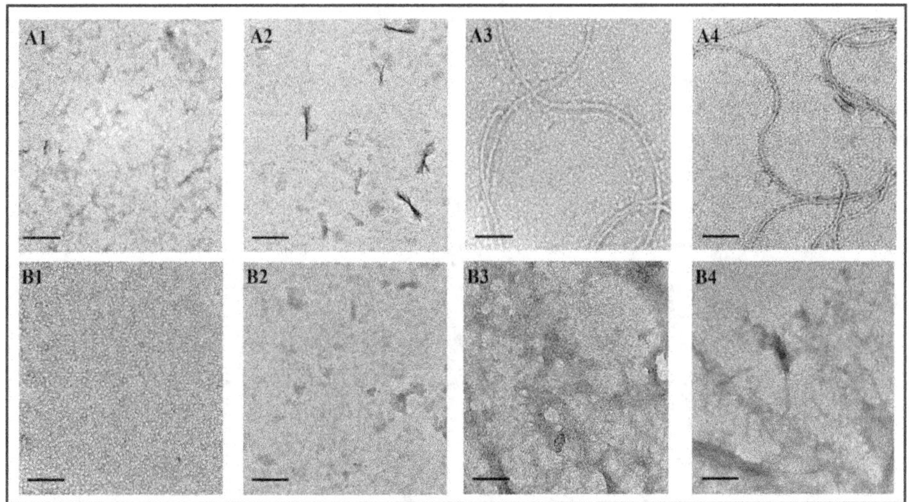

Figure 4.3. Electron micrograph of HSA fibrillated (A) alone; (B) in presence of QR (6 μM) sampled and recorded at (1) 1 hour (2) 4 hours (3) 6 hours (4) 12 hours. Scale 0.2 μm.

4.3.3. Effects of QR on matured fibrils

Matured HSA fibrils were incubated with different concentrations (0-30 μM) of QR at 37 ± 2 °C for 48 hours and the effects were evaluated through ThT fluorescence spectroscopy. Pilot experiments were also run to find the optimal incubation time effect and that was standardized. Samples when incubated with QR beyond 48 hours did not present appreciable changes in fluorescence responses. The ThT emission intensity of fibrils co-incubated with QR decreased progressively with increase in QR concentrations (figure 4.4A). The maximum fall in ThT intensity was 29 % which occurred in presence of 20 μM QR ($P<0.05$). No change in ThT intensity were detected in samples containing higher concentrations of QR.

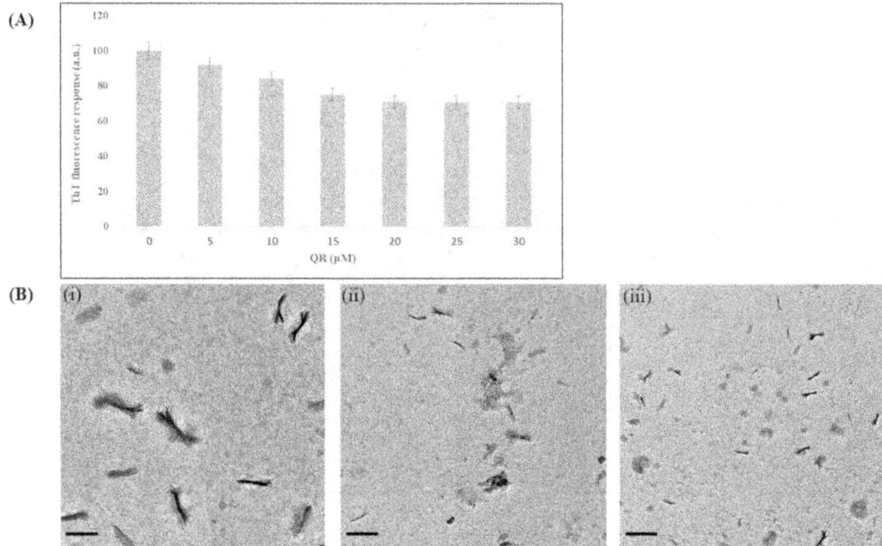

Figure 4.4. Effects of QR on preformed fibrils (A) ThT responses of HSA fibrils incubated for 48 hours with (0-30 μM) QR. ThT responses were expressed as mean±S.D. (n=3). (B) TEM studies of HSA fibrils incubated with 5, 10, 20 μM QR. Scale 0.2 μm.

Electron micrographs of the fibrils incubated with QR were also recorded. Effective disruption of fibrillar networks and presence of no fibrillar aggregates were evidenced (figure 4.4B). The protein structures appeared similar to those reported for similar other proteins fibrillar breakdown [60,61]. A decrease in the population of fibrils with increasing QR concentrations established a concentration dependant penetration and efficacy. Electron microscopy, along with ThT fluorescence spectroscopic data thus supported the disruptive effect of QR on preformed HSA fibrils. It seemed that QR exhibits site specific binding with the serum albumin protein in the native as well as non-native forms [62–64]. This apparently prevented stress induced self-association, nucleation as well oligomeric formations due to interactions with the non-native monomeric species.

4.3.4. Molecular interaction study of HSA and QR

Molecular interactions of the bioactive and the native protein was considered paramount to gain insight into the inhibitory and disruptive mechanism of QR. Diffusive interactions of HSA and QR were explored through UV absorption spectroscopy at 37±2 °C. Absorption spectra of HSA in presence of different concentrations of QR (0-4 μM) were obtained (figure 4.5A). QR did not respond in range of 250- 300 nm while that appeared strongly in case of HSA.

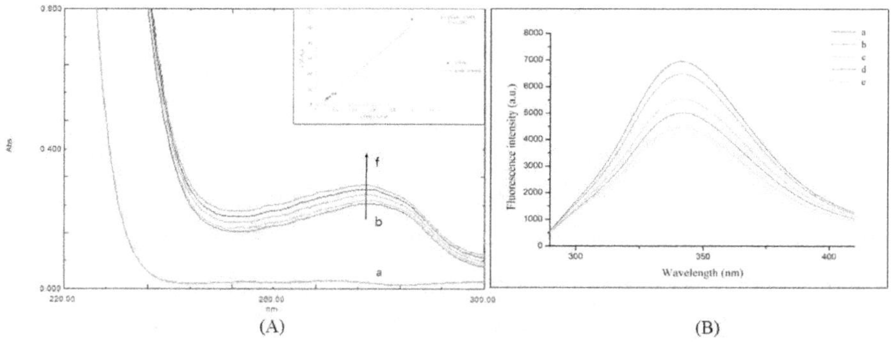

Figure 4.5. (A) UV absorption spectra of QR (a) and HSA in different concentrations of QR (0-4 μM) (b–f). Inset: Benesi and Hildebrand plot for interaction studies. (B) Fluorescence emission spectra of HSA in presence of increasing concentrations (0-4 μM) of QR (a-e).

An increase in the absorbance at 278 nm was observed with the increasing concentrations of QR. This was logically due to QR affinity diffusion and binding with the HSA motifs [65]. The non-covalent interactions appeared obvious. The value of binding constant (K) in complex formation was derived from the absorption data as described earlier in section 3.3.3. Concentration of QR was expressed in mol/l and was taken as [QR]. The binding constant was found to be 4.81×10^4 M^{-1} from the graphical plot (figure 4.5A inset). Free energy of interaction (ΔG) as calculated as -6.6 kcal/mol at 37±2 °C and that established a spontaneity in QR bioactive-protein homing.

Fluorescence emission spectroscopy was yet another reliable tool to help understand interactions of proteins and small molecules. The intrinsic fluorescence of HSA was mainly contributed by the tryptophan residue, phenylalanine has a low quantum yield and the fluorescence response of tyrosine was quenched upon ionization in medium (figure 4.5B) [66]. Increasing diffusion of QR to protein sites resulted in a progressive quenching of fluorescence emission of HSA at 350 nm. This suggested that the binding region of QR was in the vicinity of tryptophan residue of HSA [67]. Quenching effects associated with fluorescence spectroscopy as well as the UV absorption spectroscopy studies have confirmed formation of a stable QR HSA complex [68]. Non-covalent interactions were likely responsible for stabilizing the native protein structure and hindered self-association of monomers, thus extending the fibrillation lag time (figure 4.2) and reducing the propensity towards fibril formation.

4.3.5. Molecular docking studies

Molecular docking is an *in silico* approach which often complements and helps reiterate the wet lab studies in a protein-ligand interaction study [69]. The best dock pose (figure 4.6) showed the Glide XP and IFD scores as -10.54 and -25094.41 kcal/mol. This suggested a high binding affinity of QR towards the protein target. The QR extended hydrogen bonding interactions with the designated amino acids, ARG 222, GLU 230, ASP 237, ASN 267 and ILE 290 of the protein. The multiple hydroxyl groups in QR scaffold were favourable to that. The docking observations indicated predominant hydrogen bonding interactions.

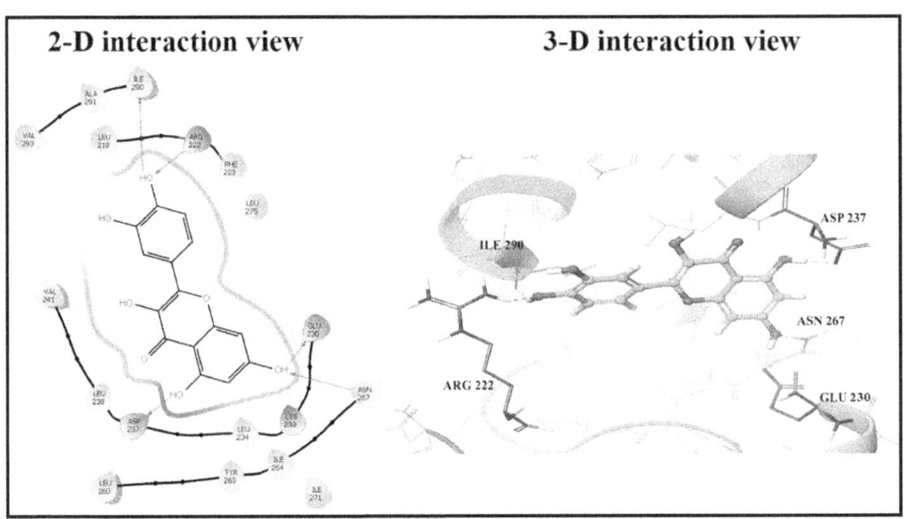

Figure 4.6. 2D and 3D studies of HSA (PDB id: 2BXP) docked to QR.

HSA fibril growth deterrence and disruptive effects due to QR flavonoid were evidenced both in the spectroscopy and electron microscopy studies. Hydrogen bonding interactions were established as one major reason responsible for shielding the active sites of amyloidogenic proteins from surface exposure of their hydrophobic residues and therefore retarding fibril formation [70,71]. The mature fibrils of HSA were stabilized by hydrophobic interactions occurring between the constituting oligomeric species [72]. QR binding to the exposed hydrophobic patches of amino acids led deceptively to the breakdown of hydrophobic forces of the oligomeric units of HSA fibrils. The mature fibrils were thus disrupted and smaller fragments were generated (figure 4.4B).

4.3.6. Cell viability studies

Quercetin is converted into an oxidation product *o*-quinone/quinone methide during scavenging of highly reactive species. This product is known to react with cell proteins and is therefore cytotoxic [73]. The cytotoxic assays of the working concentrations of QR in HaCaT cell lines were aimed to study the compatibility of the bioactive. QR solutions prepared in concentrations of 5, 10, 20, 40 and 50 μM did not elicit significant cytotoxicity upon co-culturing with the HaCaT cells (figure 4.7). Therefore the safety in application of quercetin in its fibril inhibitory and fibril disruptive concentrations was substantiated.

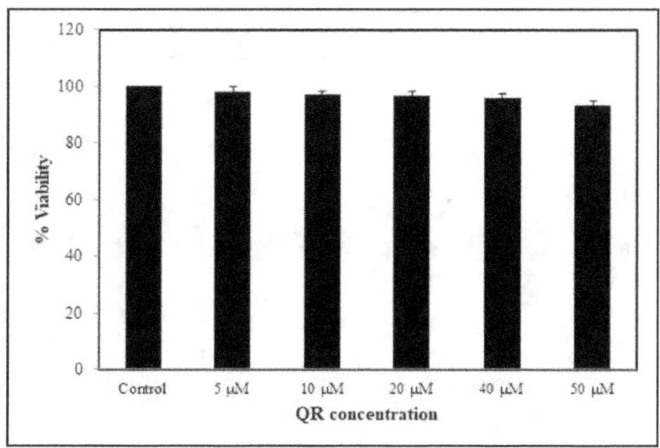

Figure 4.7. Cytotoxic assay of QR (0-50 μM) in HaCaT cell lines. Results were expressed as mean±S.D. (n=3).

4.4. Summary

QR appeared as one of the most efficient inhibitors of protein fibrillation. Non-covalent interactions like multi-level hydrogen bonding between HSA and QR appeared as one deterrent to the heat stress induced protein fibrillation. QR exhibited a site specific affinity to serum albumin and such effects were concentration dependent. QR self-interactions delayed nucleation, oligomer association and that considerably extended HSA fibrillation lag phase. A set of biophysical experiments further demonstrated the unique fibril disruptive capacity of the bioactive QR. The present study therefore highlighted that the flavonoid QR effectively inhibited the progression of HSA fibrillation and disrupted the preformed fibrils successfully.

CHAPTER 5

Effects of Silybin on Protein Fibrillation

5.1. Introduction

Silybin (or silibinin) is a bioactive flavonolignan and constitutes 34 % by mass of silymarin. Silymarin is a flavonoidic mixture obtained from the flowers and leaves of milk thistle (*Silybum marianum*) plant. The milk thistle plant and silymarin have been used by different civilizations as a natural remedy for centuries, especially in Amatoxin mushroom (*Amanita phalloides*) poisoning [74,75]. Renewed scientific interests in plant derived micronutrients in the past decade led to a number of publications related to silymarin, with over 2670 citations in 2019 [76]. The standardized mixture of silymarin containing silybin, isosilybin, silydianin, silychristin and others, is now a part of contemporary medicine available for treatment of several diseases [77–79].

Silybin (SB) has attracted considerable research interests owing to its potent anti-oxidant [80], anti-inflammatory [81], anti-diabetic [82], as well as neuroprotective properties [83]. This part of the work has been designed to investigate the effects of different concentrations of flavonolignan SB on protein fibrillation. Underlying molecular interactions between SB and HSA were also explored through spectroscopy and *in silico* docking studies.

5.2. Experimental

5.2.1. Protein fibril formation

Protein stock solution was prepared by dissolving lyophilized HSA in PBS, and its concentration (100 µM) was determined (section 3.2.1). HSA solutions were incubated with different molar ratios of SB (HSA:SB :: 1:0.25, 1:0.5, 1:1 and 1:1.5) at 65±2 °C in orbital shaker incubator, set at 80 rpm. Aliquots were diluted and effects of SB on HSA fibrillation were investigated.

5.2.2. Dynamic light scattering studies

HSA fibrillation samples (5.2.1) were diluted to 2 µM in particle-free water and average sizes were recorded on a zetasizer (section 3.2.2). Hydrodynamic radii were presented with respect to time (figure 5.1).

5.2.3. Thioflavin T fluorescence spectroscopy

HSA samples (2 µM) were allowed to incubate with ThT dye (10 µM) solution for 20 minutes [16]. The fluorescence intensities were then recorded in a spectrofluorimeter (figure

5.2), having excitation and emission wavelengths fixed at 450 and 480 nm (section 3.2.3). SB was also checked to confirm non-responsiveness to ThT fluorescence.

5.2.4. Cytotoxicity evaluations

In vitro cytotoxicity of SB was evaluated through trypan blue exclusion assay following the described protocol (section 3.2.4). Cultured HaCaT cells were either left untreated or treated with different doses of SB (5-50 µM) for 24 hours. Then the cells were harvested and trypan blue solution was added to the cell suspensions. The number of total and dead cells was counted and the averaged data derived from a set of three replicates were recoded for analysis.

5.2.5. HSA Silybin interaction studies

Silybin (SB, 24.1 mg, Sigma Aldrich, US) was dissolved in minimum volume of DMSO and volume adjusted to 10 ml with phosphate buffer so as to make 5 mM stock solution. HSA solutions (5 µM) were incubated with different concentrations of SB (0-4 µM) at 37±2 °C for 30 minutes. Absorption spectra in a range of 220 nm to 350 nm were collected on a UV-Visible spectrophotometer (section 3.2.5).

5.2.6. Molecular modelling studies

The structure of SB was prepared [18] and IFD protocol was implemented following the same three steps as described in section 3.2.6. The IFD score was generated [84], as the ligand was re-docked into refined low energy induced fit structure. IFD scores were ranked for best post selection and analysis. All molecular interactions were visualized using 2D interaction diagram utility tool and Schrödinger suite in Maestro user interface (Schrödinger, LLC, New York, NY. 2013).

5.2.7. Statistical analysis

All the experiments were performed in triplicates and the data have been presented as mean±S.D. Student's t-test was done and differences were considered significant if $P<0.05$.

5.3. Results and discussion

5.3.1. Effect of SB on HSA fibril formation

SB was incubated in different stoichiometric ratios with HSA at 65±2 °C for 12 hours and preliminary information on fibril formations were obtained through DLS experiments. Hydrodynamic radii of HSA samples in absence of SB continued to increase over time, and that was due to formation of higher molecular weight aggregates [55]. Presence of SB in ratios (HSA:SB) of 1:0.25, 1:0.5, 1:1, 1:1.5 restricted the growth of HSA aggregates concentration dependently to 181±7.2 nm, 157±6.4 nm, 135±5.7 nm, and 110±5.0 nm within 4 hours of incubation (figure 5.1). HSA aggregates in absence of SB however attained equilibrium after 6 hours of incubation. SB therefore appeared to have delayed the growth of protein aggregates. SB exhibited a concentration dependent repressive effect on the changes of HSA hydrodynamic radii upon incubation. Changes in DLS size observations were but inconsistent beyond 1:1.5.

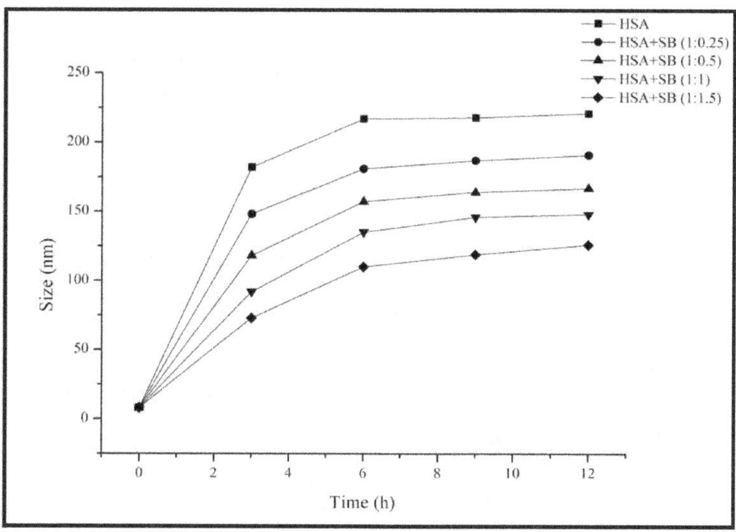

Figure 5.1. Hydrodynamic radii of HSA incubated with or without different ratios of SB at 65±2 °C.

Thioflavin T is an amyloid specific dye which produces significant fluorescence quantum yield upon binding to the cross β sheet pockets of amyloid fibrils [55]. Native HSA did not exhibit any ThT response at 480 nm due to its un-aggregated state. Increase of ThT response upon incubation of HSA at 65±2 °C demonstrated the formation of cross β sheets through progressive exposure of hydrophobic groups of native albumin structure [85]. Co-incubation with increasing ratios of SB, however, led to lowering of the ThT intensities. A mild reduction of 8 % intensity at 12 hours was observed at low HSA:SB ratio of 1: 0.25, while ThT response for the HSA samples decreased maximum by 36 % at 1: 1.5 (figure 5.2). Concentration dependent inhibitory effect of SB on HSA fibril formation at high temperature incubation was established.

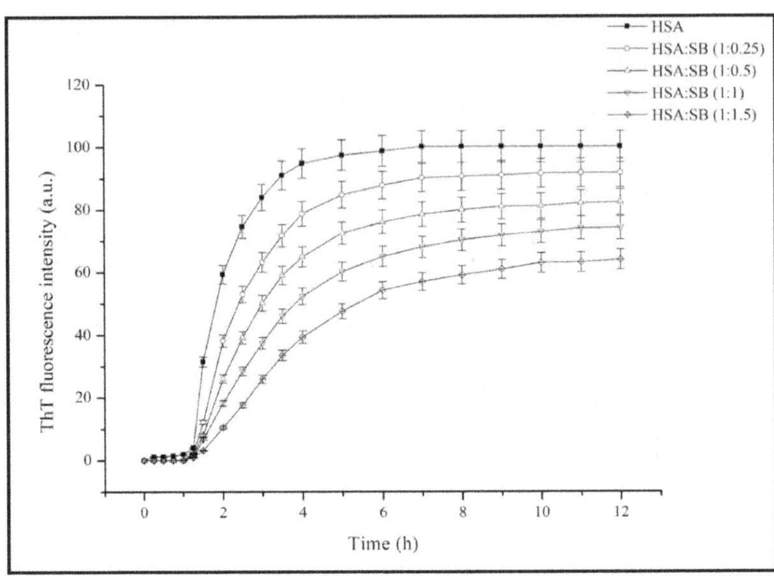

Figure 5.2. ThT fluorescence spectra of HSA in presence of different ratios of SB. ThT intensities were expressed as mean±S.D. (n=3).

Data obtained from the ThT fluorescence experiments were further fitted into equation 3.1 to interpret the lag time information.

$$F = F_{min} + \frac{F_{max}}{1+e^{-[\frac{t-t_0}{\tau}]}}$$
(3.1)

The course of HSA fibril formations at 65±2 °C followed a typical nucleation dependent polymerization model, with a lag time of 0.6 hour and the process achieved equilibrium at about 6 hours from initiation [58]. Increase of the lag phase ($P<0.05$) was apparent in presence of SB (table 5.1). SB likely repressed the misfolding of albumin and subsequently prevented the formation of nucleus, thus delaying the lag phase.

Table 5.1: Lag time and apparent rate constant of HSA fibrillation

HSA:SB	Lag time (hour)	Apparent rate constant (hour⁻¹)
1:0	0.68 ±0.03	2.68
1:0.25	1.05 ±0.11	1.86
1:0.5	1.38 ±0.17	1.45
1:1	1.77 ±0.22	1.13
1:1.5	2.19 ±0.28	1.07

Results were expressed as mean±S.D. (n=3).

5.3.2. Cell viability analysis

SB at high (>100 μM) concentrations exerted cytotoxic effects on different cell lines [86–88]. Working concentrations for SB for protein fibrillation experiments were initially preferred for lower concentration ranges. Cell viability studies showed that SB did not cause significant death of HaCaT cells at 5 – 50 μM concentrations (figure 5.3).

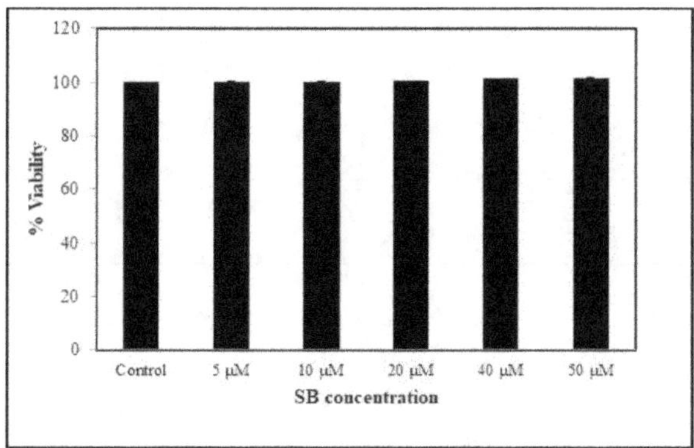

Figure 5.3. Cell viability assay of different concentrations of SB on HaCaT cell lines. Results were expressed as mean±S.D. (n=3).

5.3.3. Characterization of binding interaction between HSA and SB

UV absorption spectroscopy presents an efficient means to understand the protein conformational changes upon binding of a small molecule. The UV absorption spectra of HSA (5 μM) in PBS (pH 7.4) revealed an intense peak at 278 nm due to π –π^* transitions in the aromatic rings of tyrosine, tryptophan and phenylalanine residues [89]. Gradual addition of SB caused enhancement of absorbance at 278 nm (figure 5.4). Due to presence of aromatic rings and hydroxyl groups in SB structure, there was possible formation of π stacking and hydrogen interactions between the bioactive and the polypeptide chains. These non-covalent interactions therefore led to a ground state complex. No shift in the absorption maxima further confirmed that SB binding with albumin did not cause significant change in the protein microenvironment. Benesi and Hildebrand equation (3.2) was applied to derive the binding constant (K) from the absorption data (figure 5.4 inset). Concentration of SB was expressed in mol/l and was taken as [SB]. The binding constant was found to be 9.56 x 10^4 M^{-1} and indicated significant binding. The free energy of interaction calculated from the binding constant (equation 3.3) was -7.06 kcal/mol. The negative value of ΔG confirmed that SB binding with HSA was spontaneous.

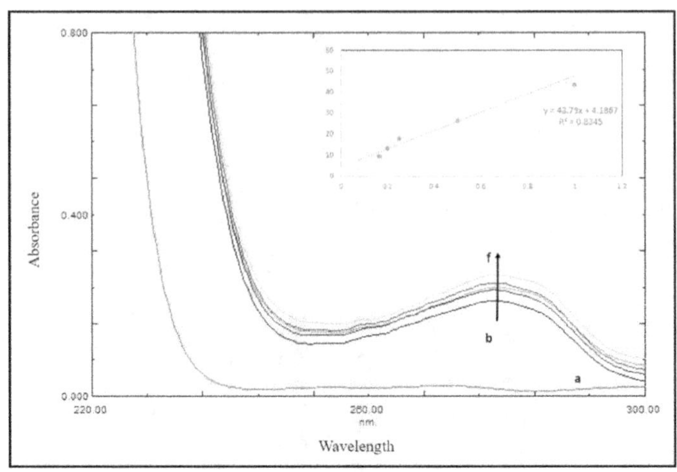

Figure 5.4. UV absorption spectra of SB (a) HSA in different concentrations of SB (0-4 µM) (b–f). Inset: Benesi and Hildebrand plot for interaction studies.

5.3.4. Molecular docking and stimulation

HSA molecular binding interactions and binding affinities for bioactive SB were investigated using extensive docking techniques. The best dock pose selection was made based on lowest docking score. The docking scores for Glide XP and IFD analyses were -10.54 kcal/mol and -2511.57 kcal/mol, respectively. In general, a high negative docking score indicates better binding affinity of any ligand towards the protein target. The binding mode of SB with HSA was represented by several numbers of hydrogen bonds and one π-cation interaction. Typical 2D and 3D binding orientation of SB has been displayed in figure 5.5. Detailed observation revealed that amino acid residues GLU 153, LYS 199, ARG 257, SER 287 and GLU 292 extended for hydrogen bond interactions formation. Among the above mentioned residues, basic amino acid ARG 257 of HSA participated to form π -cation interaction with SB. Mostly, the hydroxyl groups which are present in the SB acted as hydrogen bond acceptors for formation of hydrogen bond with HSA. All hydrogen bond interactions were observed for SB and HSA within a distance of 1.77 to 2.19 Å. Apart from hydrogen bond interactions, several hydrophobic residues such as LEU 219, PHE 223, LEU 234, ALA 258, LEU 260, ALA 261, ILE 264, ILE 290 and ALA 291 were also found in close proximity and might be accounted for exposing hydrophobic contacts with SB.

It has been established that the protein fibril structure are well stabilized by hydrophobic interactions occurring between the aromatic groups of different amino acids [90]. The *in silico* studies along with the spectroscopic data demonstrated that several non-covalent interactions were responsible for formation of HSA- SB complex. These interactions however did not cause alteration in the protein conformation and perhaps helped maintain its native form in fibrillating conditions.

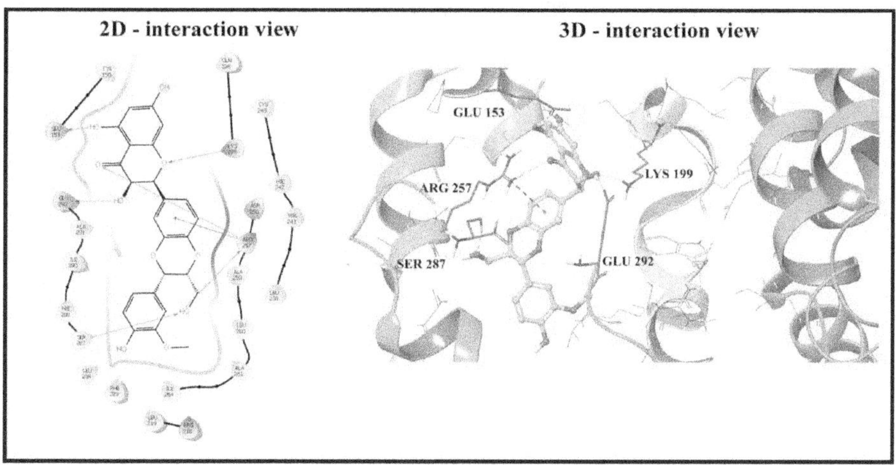

Figure 5.5. 2D and 3D view of crystal structure of human serum albumin (PDB id: 2BXP) docked to SB.

5.4. Summary

Inhibitory capacity of flavonolignan silybin against HSA fibril formation was explored and established. Interactions between SB and HSA predicted through both *in silico* and spectroscopy studies revealed the SB binds with the protein through several hydrogen bonds and hydrophobic forces. A combination of different biophysical techniques revealed that ~36 % inhibition of HSA fibril formation occurred at a HSA:SB dose ratio of 1:1.5. Cytotoxicity evaluations on HaCaT cells confirmed fibril inhibitory effects of SB at safe doses. The work demonstrated that HSA structural stabilization through non-covalent binding of silybin was the guiding factor for restricting the formation of protein fibrils at elevated temperature.

CHAPTER 6

Molecularly Tethered Nano-Gold on Protein Fibrillation

6.1. Introduction

Gold nanoparticles are veritable tools in disease diagnosis and therapy [91]. Surface chemistry in nanoscale and the particle aspect ratio can influence molecular interactions in bio-physical interfaces. Gold nanoparticles near spheres are apparently safe and elicit low or no systemic immunity reactions. Gold nanoparticles are also known to diffuse through various physiological barriers and bind to the protein macromolecules owing to particle advantages [92,93]. Furthermore, quasi-crystalline gold nanoparticles are highly porous in nature and that facilitates bioactive payload transport deeper into the macromolecular receptor targets [94,95]. Functionalized nanoparticles by design were projected as some of the successful strategies in hereto incurable disease conditions [96–98].

Polymeric nanoparticles were demonstrated earlier as profound inhibitors of protein fibrillation [99,100]. Gold nanoparticles besides being biochemically innocuous extend further advantages like tenability of shape and size, functionalization at the surfaces, payload carrier capacity, and molecular interactions.

This part of the work attempts to understand the effects of selective plant bioactives tethered gold nanoparticles on fibrillation of HSA. Fibrillation inhibition capacities of the nanoparticles were studied through a combination of spectroscopy and microscopic techniques. Gold nanoparticles were separately checked for indications of cytotoxicity at fibrillation inhibitory concentrations. Comparative studies were undertaken in order to identify potentials of bioactive tethered gold nanoparticles in protein fibrillation inhibition.

6.2. Experimental

6.2.1. Synthesis of gold nanoparticles and characterization

Polyphenols tethered gold nanoparticles were prepared in a sonicator bath set at 50 kHz (UD 100SH-3L, Takashi, Japan). Gold nanoparticles (GAuNPs, QAuNPs and SAuNPs) were synthesized at lower temperature conditions. Typically, 3 ml of 1.3 mM chloroauric acid solution was allowed to react in batches, under sonication, with the corresponding reagents, gallic acid (GA), quercetin (QR) or silybin (SB). Samples were withdrawn at different time intervals and studied in a UV-Visible spectrophotometer to understand the appearance and optimization of particle plasmon response. Reagent concentrations were generalized at 4 mM, temperature was meticulously maintained at 4±2 °C and a reaction time of 1 minute was devised in case of GA and QR. SB having lower oxidative potential required a reaction temperature of 45±2 °C and optimal time of 30 minutes. As synthesized nanoparticles were preliminarily centrifuged (C-30, Remi, India) at 6,000 rpm, 4±2 °C for 10 minutes and sediments, if any, were discarded. Supernatants were collected, re-centrifuged at 14,500 rpm, for 60 minutes at 4±2 °C and nano-gold pellets were collected. The pellets were re-dispersed in water, re-centrifuged and washed twice before being collected in water dispersions for storing in refrigerators.

Surface plasmon response of nanoparticles (GAuNPs, QAuNPs and SAuNPs) were recorded using a double beam UV-Visible spectrophotometer, equipped with matching quartz cuvettes of 1 cm path length. Sample spectra were obtained at medium scanning speed in a wavelength range of 200–800 nm. Hydrodynamic diameters of samples were obtained from a zetasizer equipped with a 4 mW He–Ne laser beam, 633 nm having a back scattering angle of 173°. Samples for electron microscopy were prepared by placing a drop of diluted sample on carbon coated copper grid (Ted Pella Inc., US). Micrographs were captured at an accelerating voltage of 200 kV using a transmission electron microscope. FT IR spectra over a range of 4000-400 cm^{-1} in pelletized samples were collected at a resolution of 4 cm^{-1} using a Jasco-670 Plus FT IR spectrometer (Jasco, Japan). The particle size results were derived from an average of 50 particles recordings in the electron micrographs and applied for the working concentration (100 nM) of gold nanoparticles in solution [101]. Concentration of gold in solution was determined using a formula:

$$C = A_{450} / (2.18 \times 10^8 \ M^{-1} \ cm^{-1}) \tag{6.1}$$

Here, C was the concentration of gold nanoparticles (in molar) and A_{450} was absorbance at 450 nm.

Free radical scavenging activity of water-dispersible nanoparticles was estimated following DPPH (2,2-diphenyl-1-picrylhydrazyl) assay [102]. Briefly, 0.5 ml of each sample was reacted with 1 ml of 0.2 mM methanolic DPPH solution and the final concentrations of nanoparticles in all cases were kept in the range of 10-100 nM. Methanol served as the negative control and ascorbic acid was used as positive standard for comparative analysis [103]. Absorbance of each sample at 517 nm was recorded after 30 minutes of incubation. The percentage of DPPH scavenging activity of gold nanoparticles was derived using the formula:

$$\text{DPPH scavenging activity} = \frac{\text{Absorbance of negative control} - \text{Absorbance of sample}}{\text{Absorbance of negative control} - \text{Absorbance of positive control}} \times 100 \tag{6.2}$$

6.2.2. Protein fibril formation

A stock solution of HSA was prepared by dissolving lyophilized HSA in PBS and the concentration was determined (as described in section 3.2.1). HSA solutions (2 µM) were incubated in presence and absence of nanoparticles at 65±2 °C for protein fibrillation studies (3.2.1) [104]. The nanoparticle: HSA ratio was maintained throughout at 1:200 [105].

6.2.3. Thioflavin T assay

Protein samples (containing 10 nM concentrations of GAuNPs, QAuNPs or SAuNPs) were withdrawn and were incubated for 20 minutes with ThT solution (20 µM) (section 3.2.3) [16]. Fluorescence intensities were recorded using a spectrofluorimeter. The excitation and emission wavelengths were set at 450 and 480 nm. The slit width used was 5 nm.

6.2.4. Cytotoxicity studies

In vitro cytotoxicity assays of gold nanoparticles were done based on trypan blue exclusion method (section 3.2.4). HaCaT cells were treated with different concentrations of gold nanoparticles ranging from 5-20 nM. The cells were left for 24 hours for treatment and were harvested for trypan blue exclusion assays. The viable cell percentages were calculated and plotted.

6.2.5. Circular dichroism

The CD spectrum in each case was recorded as described earlier (section 3.2.8). Scan speed was set as 100 nm/minute. The spectral data were processed using an online server BeStSel (http://bestsel.elte.hu) and the secondary structural estimates were recorded.

6.2.6. Fluorescence microscopy

HSA samples (each 15 µl) incubated with or without gold nanoparticles were stained with 1mM ThT solution (15 µl) and applied onto glass slides. Micrographs were recorded using Axiovert 40 CFL (Carl Zeiss, Germany) microscope.

6.2.7. Transmission electron microscopy

Highly diluted samples were placed on carbon coated copper grids, negatively stained with 2 % w/v uranyl acetate solution and electron micrographs were recorded.

6.2.8. Statistical analysis

Most experiments were performed in triplicates unless mentioned otherwise. The results were presented as mean± standard deviation (S.D.). Origin 6.0 Professional was used for the analysis of the experimental data. Student t-test was conducted for comparative studies and the difference was considered significant when $P< 0.05$.

6.3. Results and discussion

6.3.1. Synthesis and characterization of gold nanoparticles

Gold nanoparticles formation in bottom up technique is a two-step process involving initial nucleation and molecular coalescence [106]. Turkevich originally described near uniform gold nanoparticles synthesis using citrate as an *in situ* reactant [107,108]. The reaction temperature used was much higher ≈100 ^{0}C though. Several modifications were later instituted but the strength of the reductant used and the action of a stabilizer in the solution phase appeared critical [109,110]. Often stabilizers such as CTAB (cetyl trimethyl ammonium bromide) were used which were known tissue-toxic agents. A solution phase synthesis was developed using ultrasound waves as one electron transfer promoter [111]. In water medium, ultrasounds create pressure waves and induce molecular vibrations. Ultrasound waves generate compression cavities inside the bulk medium. Short lived water bubbles develop due to cavitational forces and crush very fast. The energy released due to cavitational collapses in bubble centre is extremely high [103,112]. Polyphenol molecules under ultrasound conditions have enabled a rapid reduction, gold nucleation and particle growth for stabilization in presence of polyphenol excesses. A generalized gold nanoparticles synthesis in plant bioactives was thus devised (scheme 6.1). The temperature and time control could apparently be used for near monodispersed gold nanoparticles synthesis using plant bioactives, GA, QR and SB. When experimented with AG, the nanoparticles appeared larger in size and were polydispersed. Unlike GA, AG is a diterpenoid lactone and is not a reactive polyphenol. GA was therefore used a simpler plant bioactive polyphenol for representative synthesis studies.

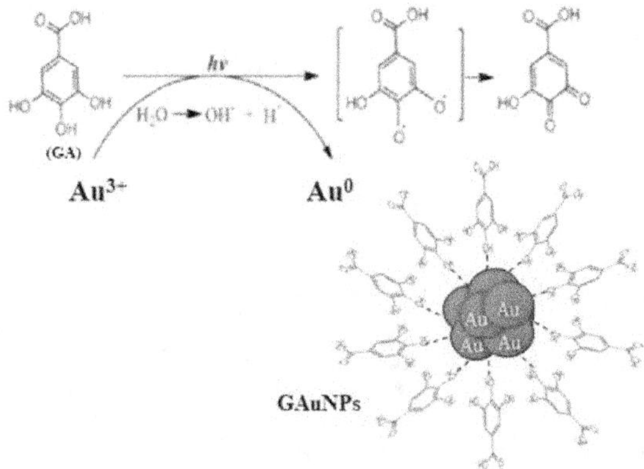

Scheme 6.1. Synthesis of gold nanoparticles.

SAuNPs were confirmed by a characteristic surface plasmon resonance (SPR) at 547 nm, and similar peaks were recorded through UV-Visible absorption spectral analysis at 541 nm and 539 nm for QAuNPs and GAuNPs (figure 6.1A). The SPR peak depends upon the shape and size of metallic nanoparticles, and on the dielectric constant of the dispersion medium [113]. The hydrodynamic diameters of gold nanoparticles were in a close range of 50 to 80 nm. High resolution electron micrographs revealed that the polyphenolic tethered gold nanoparticles were near spherical in shape. TEM studies further confirmed that the average particle diameters of SAuNPs, GAuNPs and QAuNPs were 15.35± 0.26 nm, 15.05± 0.20 nm and 15.12± 0.21 nm, respectively (figure 6.1D). Particle size is amongst the key parameters which guide the association of protein molecules on nano-materials [114]. Near monodispersity of nanoparticles were revealed in both DLS and microscopy studies.

FT IR studies were undertaken to understand presence of various polyphenols on the gold nanoparticle surfaces (figure 6.1C). FT IR spectra revealed bands at 3420 cm^{-1}, 1638 cm^{-1} and 1508 cm^{-1}, arising from phenolic –OH groups, C=O stretching and aromatic ring stretching [111,115]. These are common facets in polyphenolics and were abundant on the gold nanosurfaces. A stretching vibration near 1088 cm^{-1} was observed in case of SAuNPs and QAuNPs, which was attributed to the benzopyran rings. Thus it appeared that phenolic –OH groups assisted electrostatic stabilization of nanoparticles in solution [113].

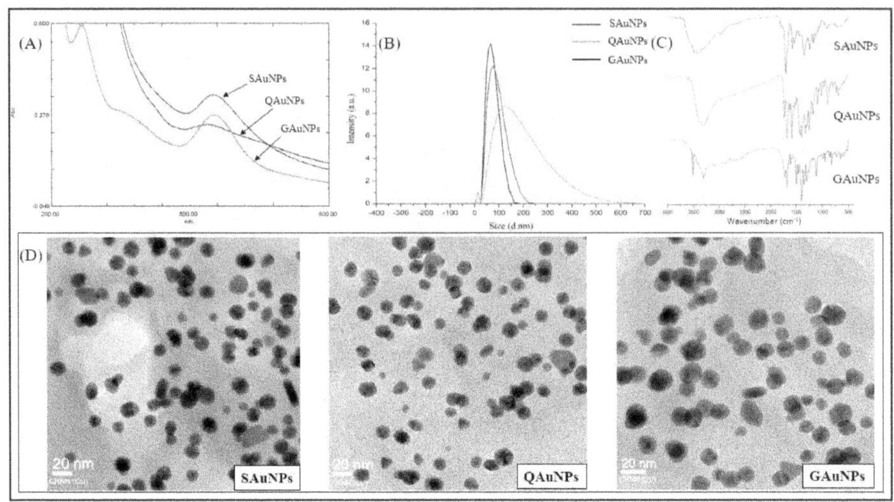

Figure 6.1. Characterization of different gold nanoparticles. (A) UV-Visible spectra; (B) Hydrodynamic diameter in DLS; (C) FT IR spectra; (D) Electron micrographs.

6.3.2. Free radical scavenging studies

Hydroxyl groups of polyphenols imparted antioxidant property to the nanoparticles. Discoloration of DPPH was used for *in vitro* estimation of antioxidant capacity [116,117]. The percentages of DPPH scavenging activities of nanoparticles were found to be a function of their concentrations. A concentration dependent linearity was observed in case of all gold nanoparticle types synthesized (for SAuNPs: R^2= 0.9989; QAuNPs: R^2= 0.9678; GAuNPs: R^2= 0.9667). Percentage scavenging activities of GAuNPs and QAuNPs were in higher ranges as compared to SAuNPs (figure 6.2). This appeared passable due to the differences in oxidative potentials of stabilizing molecules.

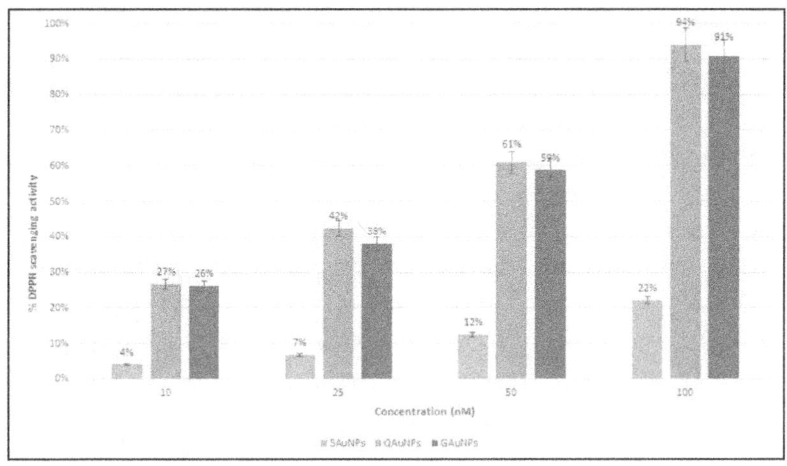

Figure 6.2. DPPH scavenging activity of SAuNPs, QAuNPs and GAuNPs at different concentrations (10-100 nM). Percentages were expressed as mean±S.D. (n=3).

6.3.3. Effect on gold nanoparticles on HSA fibrillation

Fibrillation kinetics of HSA in presence of gold nanoparticles have been studied intently through ThT fluorescence spectroscopy (figure 6.3). The ThT intensities of HSA samples withdrawn at different time points continually increased in a sigmoidal manner and were similar to previous observations (section 3.3.1). Maximum ThT intensity recorded from matured HSA fibrils was used to normalize ThT fluorescence data. Citrate capped gold nanoparticles commonly synthesized following Turkevich process remained abortive to elicit inhibitory effects on HSA fibrillation upon incubation over a period of 24 hours. However, gold nanoparticles surface chirality and the presence of functional groups could inflict differential inhibition of HSA fibrillation [118]. Arguably, this was due to molecular orientation coupled with access limitations to the crucial nucleation sites of protein.

The HSA fibril growth curve considerably deviated when incubated in presence of SAuNPs. ThT data were fitted into a sigmoidal equation (equation 3.1) to derive the lag time information. On achieving steady state, SAuNPs co-incubated HSA exhibited 15.2 % decrease in ThT intensities. The responses from QAuNPs containing HSA solution however reduced greatly throughout the incubation period ($P<0.05$). No significant change in lag phase was observed upon fibril growth kinetics analysis, although ThT value at equilibrium was diminished by 67.6 %.

Similar effects on the lag phase of HSA were recorded when GAuNPs were co-incubated in identical conditions. A decrease in ThT intensity by 60.2 % detected after 24 hours of incubation indicated that GAuNPs hindered formation of HSA fibrils similarly. It was inferred that though presence of different gold nanoparticles in HSA solution did not impede nucleation of protein monomers, a diminution in fibril content was but observed in the elongation and saturation stages.

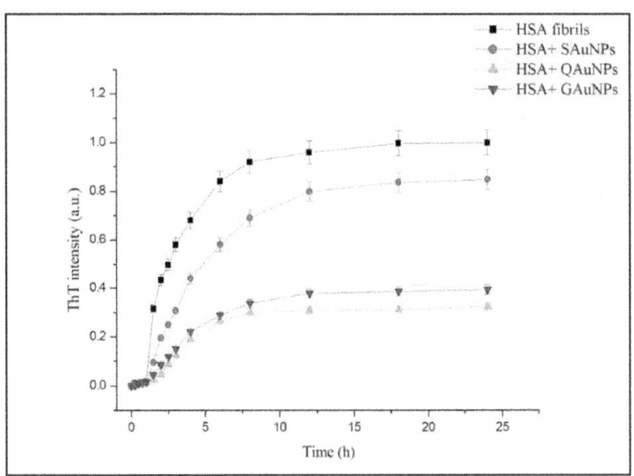

Figure 6.3. ThT responses of HSA incubated in absence and presence of gold nanoparticles. ThT intensities were expressed as mean±S.D. (n=3).

6.3.4. Effect on gold nanoparticles on protein secondary conformation

The CD spectra of native HSA and fibrils in presence and absence of gold nanoparticles were presented in figure 6.4(a). HSA solution at pH 7.4 exhibited the two negative peaks at 208 and 222 nm, owing to α helical structure (section 3.3.5). Samples when incubated for fibrillation studies showed disappearance of negativity at 208 and 222 nm due to loss of α helicity. Appearance of a single negative peak at 215 nm denoted an increase of β sheet contents, which ultimately led to protein fibrillation [85,119]. Incubation with SAuNPs effected a significant change in the CD spectrum. Higher negative mdeg values were recorded in case of GAuNPs and QAuNPs. It appeared that the GAuNPs and QAuNPs when in contact could effectively retain α helical features of HSA. The histogram in figure 6.4(b) indicated that although native HSA contained ~72.4 % of α helical structures and ~8.5 % β sheets; the same samples acquired after 24 hours fibrillation exposure exhibited ~45.8 % α helical and 24.1 % β sheet contents. In presence of SAuNPs, α helical contents of HSA fibril were ~49 %, but changes in the β sheet content appeared minimum. However, it was notable that GAuNPs when co-incubated with HSA expressed lower α helix contents ~56 % and β sheet contents ~16 %. QAuNPs effects were further prominent, where HSA α helix contents were ~61.2 % and β sheet contents were only ~14 %. The α helix structural retention therefore appeared most pronounced in case of QAuNPs.

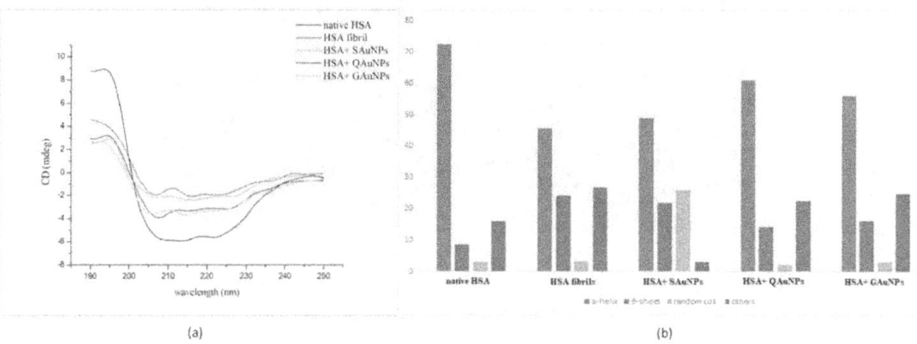

Figure 6.4. (A) CD spectra and (B) Histogram of secondary structural contents of HSA incubated in absence and presence of gold nanoparticles.

6.3.5. Morphological analysis of aggregated HSA

The formation and distribution of fluorescent dye stained HSA fibrils were monitored through fluorescence microscopy. Gold nanoparticles expressed a distinctive UV-Visible response and fluorescence microscopy study was considered to present a distinguishing effect in case of fibrils. Standalone HSA fibrils appeared as irregular patches with high fluorescent intensities scattering throughout (figure 6.5A). The number of fluorescent patches diminished after incubation with the gold nanoparticles. The fluorescence intensity and patches were significantly lower in case of SAuNPs co-incubated samples (figure 6.5B). The fluorescent patches were but almost disappeared where HSA was incubated with QAuNPs or GAuNPs (figure 6.5C, D).

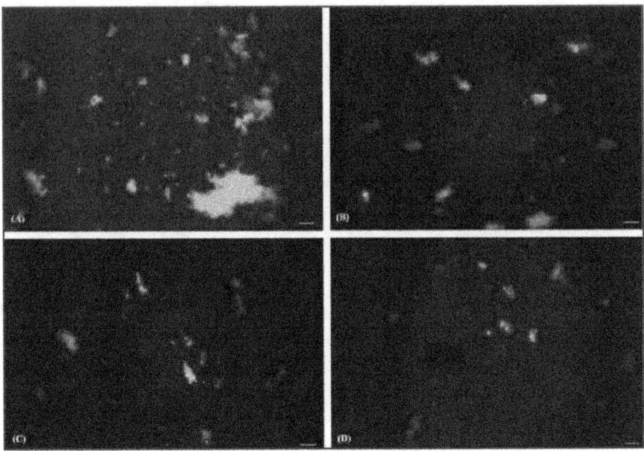

Figure 6.5. Fluorescence micrographs of HSA fibrils (A) standalone, (B) incubated with SAuNPs, (C) incubated with QAuNPs, (D) incubated with GAuNPs. Scale bars 100 μm.

Figure 6.6 showed the electron micrographs of HSA incubated in absence or presence of differently tethered gold nanoparticles. HSA fibrils showed a network of entangled fibrils [120–122]. HSA co-incubation with SAuNPs caused protein adsorption on the nano-surfaces and formation of disorganised network of protein aggregates. Decrease in fibrillar structures and presence of protein aggregates on gold nanosurface were noted as well in case of QAuNPs and GAuNPs co-incubated HSA samples. It appeared that HSA co-incubated with QAuNPs contained significantly lesser aggregates in comparison to others. Morphological observations in TEM corroborated well with the spectroscopy observations.

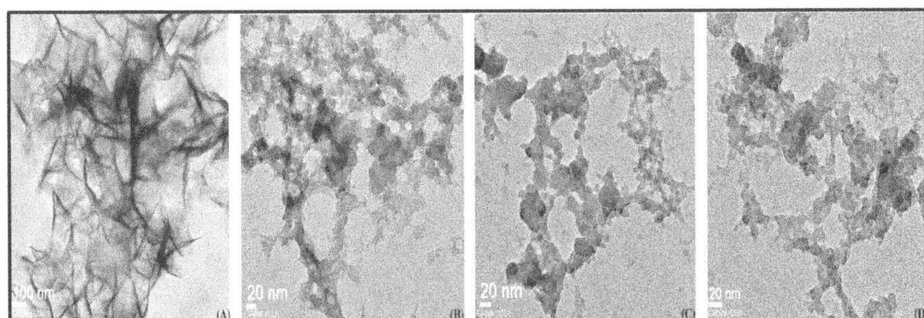

Figure 6.6. Electron micrographs of HSA fibrils (A) standalone, (B) incubated with SAuNPs, (C) incubated with QAuNPs, (D) incubated with GAuNPs.

6.3.6. Cell viability studies

Gold nanoparticles are though very useful tools in disease diagnosis and therapy, cytotoxicity remains a catching issue. Besides understanding the fibrillation inhibition efficiency of the gold nanoparticles, the cell viability studies were important for selection of an appropriate inhibitor [105]. Nanoparticle cytotoxicity analysis (figure 6.7) on HaCaT cell line revealed no significant cell death for samples containing varying concentrations of gold nanoparticles (5-20 nM). Plant bioactives capped gold nanoparticles thus appeared as biosafe in cell surface contacts. The presence of antioxidant polyphenolics on nano-gold surfaces appeared as one important contributor to its safety [123,124].

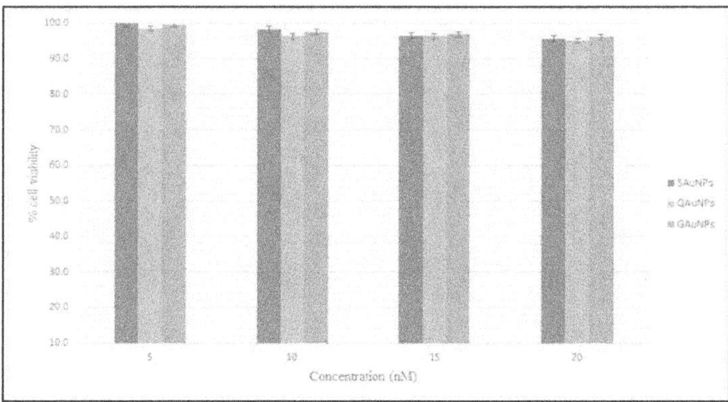

Figure 6.7. Percentage cell viability of gold nanoparticles at different concentrations (5-20 nM). Results were expressed as mean±S.D. (n=3).

6.4. Summary

Selective plant bioactives were used for *in situ* synthesis and molecular tethering of biosafe gold nanoparticles. Different gold nanoparticle types (GAuNPs, QAuNPs and SAuNPs) were synthesized following a generalised synthesis technique adopted under low temperature sonication. Nanoparticles harvested were near spherical and were uniform over a narrow size range. Gold nanoparticles molecular tethering and antioxidant capacity were analysed. The QAuNPs expressed significant anti-oxidative capacity in DPPH assay. Effects of different gold nanoparticles in HSA protein fibrillation were studied in depth. Gold nanoparticles appeared as efficient inhibitors of protein fibrillation. That effect was attributed to nanoparticles' Brownian motion, rapid diffusion in protein interfaces and surface functionalization. Relative efficiency of gold nanoparticle types in inhibition of protein fibrillation could easily be correlated with particle antioxidant capacity. Apparently, gold nanoparticles were efficient carriers of plant bioactives. Functionalized gold nanoparticles extended a major potential as inhibitors of protein fibrillation.

CHAPTER 7

Effects of Biopolymer Nanochemistry in Protein Interfaces

7.1. Introduction

Biopolymer nanoparticles have been extensively experimented for encapsulation of various classes of bioactives [125]. They can conveniently degrade in physiological environment which both reduces the toxicity and help release of payload [126]. Poly (lactic-*co*-glycolic acid) is a common biopolymer used for synthesis of nano-scale materials, and can be tuned to serve specific requirements. Functionalization of PLGA nanoparticles with site targeting ligands, polyelectrolyte depositions to enhance cellular uptake or PEGlyation to diminish systemic elimination are some common strategies to design smart drug carrier devices [127,128].

Nanoparticles have been developed for various therapeutic interventions, such as pin pointed targeting of bioactives, high cellular uptake, and controlled release of payload [129,130]. Despite several advances in preclinical development of nanoparticle drug delivery strategies, there has been a therapeutic shortfall upon exposure of nanoparticles in the systemic circulation. A survey revealed that only 0.7 % of administered dose of nanoparticles are effectively delivered to its intended target site [131]. When a nanoparticle enters a biological environment, plasma proteins in the immediate vicinity get rapidly adsorbed on the nanosurface. This compromises the surface structures of the nanoparticle at molecular scale as well as alters its biological identity. The resulting form, termed as "protein corona", influences the cellular uptake and subsequent fate of the nanoparticle *in vivo* [132,133]. Understanding the facets of the protein corona is therefore essential for the progression from *in vitro* to *in vivo* experiments, which would ensure successful translation to clinical applications.

This work has been designed to investigate the effects of surface chemistry of PLGA nanoparticles on their hard corona formation. At first, pristine Pluronics stabilized and polyelectrolytes such as chitosan and heparin coated PLGA nanoparticles of a uniform size range were synthesized. The hard corona surrounding the nanoparticles and their molecular interactions formed the core of this study. Further experiments to identify the effects of these interactions on protein secondary structures were carried out. This work was therefore intended to develop a better understanding of the influence of nanosurface chemistry on protein corona formations, which would further help to design nano-carriers for *in vivo* applications.

7.2. Experimental

7.2.1. Preparation of nanoparticles and characterization

PLGA nanoparticles were prepared through a solvent diffusion method [115]. Biopolymer PLGA 50:50, 50 mg, (M.W. 40,000-75,000, Sigma Aldrich, US) was dissolved in 5 ml of acetone, and injected (2 ml/min) into 10 ml aqueous solution of 1 % w/v Pluronic F-127 (Sigma Aldrich, US) under continuous stirring at room temperature (25±2 °C). Nanoparticles (Plgn) formed were harvested after 3 hours of stirring by centrifugation at 16,000 rpm for 45 minutes. Plgn were re-dispersed in 10 ml of water for polyelectrolyte coating. Chitosan (M.W. medium, Sigma Aldrich, US) or heparin (M.W. 17,000-19,000, Sigma Aldrich, US) were dissolved in water (0.05 % w/v) and the pH of solution was kept at 6 (NaOH/HCl). Plgn nanosuspension (3 % w/v, 2 ml) was then injected into a 10 ml polyelectrolyte solution under continuous stirring. Solution was equilibrated for 5 minutes more and the resultant polyelectrolyte surface coated nanoparticles (PlgnC or PlgnH) were collected through centrifugation at 16,000 rpm followed by a washing cycle.

Gravity-driven sedimentation cycle was further performed to separate the nanoparticles in a desired size range [134]. Briefly, nanoparticle suspensions were subjected to centrifugation at 6,000 rpm for 10 minutes to sediment any large aggregates. The supernatant was then carefully collected and re-centrifuged at 14,500 rpm. The sediment particles were dispersed in 10 ml of water and the stocks were kept at 4±0.5 °C for further experiments. Samples from preparation batches were also lyophilized (FDU 1200, Eyela, Japan) for solid state studies.

Particle size analysis of diluted samples was performed in a zetasizer following the procedure described in section 6.2.1. Zeta potentials in water were also recorded on the basis of electrophoretic mobility for comparative analysis. For TEM analysis, diluted samples deposited on carbon-coated copper grids were stained with uranyl acetate, and micrographs of nanoparticles were collected.

Chitosan or heparin deposition on nanoparticles were estimated through colorimetric assay methods. Alizarin red dye (0.1 % w/v) interactions with chitosan standard solutions (x; 10- 100 µg/ml) at 571 nm [7] were applied and absorbance (y) were recorded to generate a standard graph, $y = 0.002x + 0.084$, $R^2 = 0.998$. Aliquots of the dye (0.8 ml) and chitosan containing samples were made up to 10 ml with sodium acetate buffer (pH 5.0) and allowed to interact for 30 minutes prior to analysis. Chitosan deposition on PlgnC was quantified from the difference of concentration between initial solution and harvested supernatants. Heparin was estimated using toluidine blue assisted colorimetric reactions [135]. 2.5 mg of toluidine blue and 100 mg of sodium chloride were dissolved in a 50 ml solution of 0.01 M hydrochloric acid to prepare toluidine blue stock solution. Standard solutions of heparin (x) were reacted with toluidine blue (0.025 % w/v) and absorbance (y) at 562 nm were used to develop a concentration dependant standard graph y= 0.038x +0.127, R^2 =0.992. Heparin attachment on nanoparticles was estimated from the concentration difference of the polyelectrolyte stock solution and in post harvested supernatants. All colorimetric experiments were run in triplicates and data were presented in percentage terms.

7.2.2. Nanoparticle protein corona

Lyophilized human plasma (Sigma Aldrich, catalogue no. S2257, US) was first dissolved in 1 ml of PBS. The reconstituted solution was diluted to 1 % v/v using PBS for use in further experiments [136]. Nanoparticles (500 µg/ml) were incubated with 1 % plasma solutions at 37±0.5 °C for 1 hour. The solutions were then centrifuged at 14,500 rpm to acquire the hard corona covered nanoparticles. Deposited pellets were dispersed in 1 ml of buffer solutions and washed by centrifugation to ensure nanoparticles recovery. Hard corona nanoparticles were re-suspended in PBS for particle measurements prior to relevant other analyses.

7.2.3. Protein quantification

Quantification of the protein on nanoparticles surfaces before and after corona formation was carried out using Bradford reagent (Sigma Aldrich, US) [137]. Typically, 1 ml of 1 % plasma was allowed to incubate with or without 500 µg of nanoparticles for 1 hour. Supernatant (0.1 ml) was harvested after centrifugation was added to 3 ml of Bradford reagent, and analysed at 595 nm (y) in a UV-Visible spectrophotometer. Concentration (x) dependant standard graph, y= 0.690x − 0.071, R^2=0.999, was developed from HSA solutions (0–1.4 mg/ml) in PBS and was used for protein estimations throughout.

7.2.4. Fluorescence spectroscopy

Intrinsic fluorescence response from 1 % v/v human plasma in PBS was recorded in a spectrofluorimeter. Different concentrations (0-40 µg/ml) of nanoparticles were incubated with plasma at 37±0.5 °C for 1 hour [138]. Emission quenching was recorded in the range of 290–400 nm when the excitation wavelength was set at 280 nm. Data gathered were plotted in a log- log graph to obtain the plasma protein affinity constant for each nanoparticle type [139].

7.2.5. FT IR spectroscopy

Spectra of nanoparticles and their protein corona forms were recorded over mid IR range of 4000-400 cm^{-1} using a FT IR instrument (section 6.2.1). Lyophilized samples were pressed in potassium bromide to form transparent pellets [140]. Data obtained were stacked using Origin 6.0 Pro software for analysis of the nanoparticles and their protein corona forms.

7.2.6. Gel electrophoresis

Proteins on biopolymer surfaces were determined by sodium dodecyl sulphate polyacrylamide gel electrophoresis (SDS PAGE) method [141]. Nanoparticle-protein coronas obtained (section 7.2.2) was re-dispersed in 20 µl of 1X loading buffer. Samples were treated at 90 °C for approximately 2 minutes and cooled to room temperature. Nanosurface bound proteins were separated on 10 % polyacrylamide gel in an electric field run under a voltage of 100 V for 45 minutes. Resulting gels were fixed by Coomassie blue dye and image was captured using a gel documentation system (BioRad, US). Densitometry studies were performed using Quantity One software (version 29.0, BioRad, US). Plasma samples and purified HSA were run along-side for comparative purposes.

7.2.7. Thioflavin T fluorescence studies

Nanoparticles (Plgn, PlgnC or PlgnH) were dispersed in 1 % plasma solution and incubated at 37±0.5 °C for 1 hour. The samples were re-incubated with ThT dye (10 µM) for 20 minutes and the fluorescence emissions at 480 nm were recorded [142]. Amyloid-like fibrillation in 1 % plasma was further induced at 60±2 °C for 1 hour and that was considered as the positive control.

7.2.8. Circular dichroism

Far-UV CD spectra of diluted samples (from section 7.2.2) were obtained on a CD spectropolarimeter as described in section 3.2.8.

7.2.9. Statistical analysis

All experiments were performed in triplicates and their data were expressed as mean±S.D. Comparative data were interpreted using student t-test and the difference was considered significant only when $P < 0.05$.

7.3. Results and discussion

7.3.1. Nanoparticle characterization

Studies on surface functionalized PLGA nanoparticles were carried out, as they attract tremendous potentials as carriers for large number of bioactives [143]. Bare Plgn prepared through solvent diffusion were also coated with chitosan or heparin. Mono-dispersed nanoparticles (figure 7.1), were used to understand the effects of nanoparticle surface functionalization on protein hard corona formations. TEM micrographs revealed that the particles were spherical, and that eliminated any effects due to nano-structural dissimilarities [144].

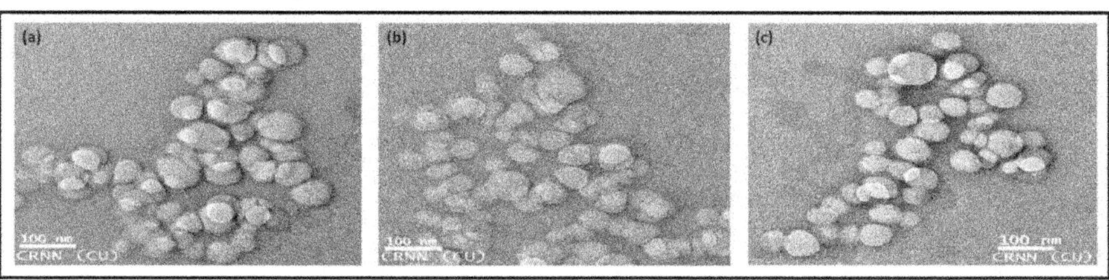

Figure 7.1. TEM micrographs of (a) Plgn (b) PlgnC and (c) PlgnH.

Characterization of nanoparticles through independent procedures confirmed polyelectrolyte deposition on biopolymer nanosurfaces. Polyelectrolyte mass on nanoparticles was estimated using alizarin red and toluidine blue assisted colorimetric reactions for chitosan and heparin, respectively. The deposition efficiency of chitosan coating was found to be 74.3±3.1 %, while the value was 51.7±2.2 % for heparin deposition. Chitosan is an amphiphilic poly-cation composed of a linear chain of β–(1-4) linked 2-acetamide-2-deoxy-D-glucose and 2-amino-2-deoxy-D-glucose units. The biopolymer forms LBL (layer-by-layer) deposition on anionic surfaces like that on Plgn. Heparin on the other hand is a hydrophilic glycosaminoglycan, composed mostly of 1→4 linked, 2-O-sulfo-α-L-iduronic acid, β-D-glucuronic acid, 2-deoxy-2-sulfamino-6-O-sulfo-α-D-glucose, 2-acetamido-2-deoxy-α-D-glucose, α-L-iduronic acid. The presence of iduronic acid influences greatly in heparin three dimension formations and self-assembly. Polymer electrostatic interactions have contributed in nanoparticle variable depositions. Both the electrostatic interactions and the polymer 3D formations have influenced in differential nanoparticle surface characteristics.

FT IR studies also confirmed chitosan and heparin molecules association onto the PLGA nanoparticles. Bare Plgn showed characteristic carbonyl peak at 1759 cm^{-1}. C-O and C-

H vibrations were recorded at 1456 cm^{-1} and 1169 cm^{-1}, respectively. Chitosan upon Plgn were identified from the presence of hydroxyl peak at 3438 cm^{-1}, and amide bands at wavenumbers 1636 cm^{-1}, and 1533 cm^{-1}. The heparin associated Plgn revealed vibrations of heparin COO$^-$ groups at 1632 cm^{-1} and a characteristic stretching band due to CH$_2$SO$_3^-$ at 959 cm^{-1}. Another wide band at 3426 cm^{-1} was observed due to presence hydroxyl moieties of heparin molecules.

The zeta potential recordings for Plgn, PlgnC and PlgnH were -19.7, +22.3 and -38.4 mV. These revealed the variations in nano-surface chemistry in each type of particles and corroborated well with the FT IR observations. Negative charge of Plgn was predominant due to exposed carboxylic groups in PLGA polymer. Chitosan deposition with exposed amide groups were responsible for positive charge of PlgnC, whereas sulphonate and carboxylate groups of heparin provided high charge densities on PlgnH surface.

7.3.2. Protein corona formation and evaluations

Formation of protein corona on a nanoparticle is a dynamic process involving reversible and irreversible attachment of protein molecules onto the surfaces [145]. Protein corona formation was initiated by incubating 500 µg of differently functionalized PLGA nanoparticles for 1 hour in PBS containing 1 % human plasma. Hard coronas of the functionalized nanoparticles collected through centrifugation, constituted of irreversibly attached proteins. DLS analysis at dilute conditions revealed the primary evidences of protein corona formation on biopolymer nanoparticles. Increase in the hydrodynamic diameters (figure 7.2a) of individual particles were significant ($P<0.05$) and that were ranging from 52.1 % (for Plgn) to 171.2 % (for PlgnC). Varying amount of plasma protein (figure 7.2b) was adsorbed on nanoparticle types. Thus nanoparticles surface functionalization was considered as one element for influencing in protein hard corona formations. Similarly, protein adsorption on the nanoparticles surfaces influenced particle aggregation and dry state surface fusion in TEM images (figure 7.2c).

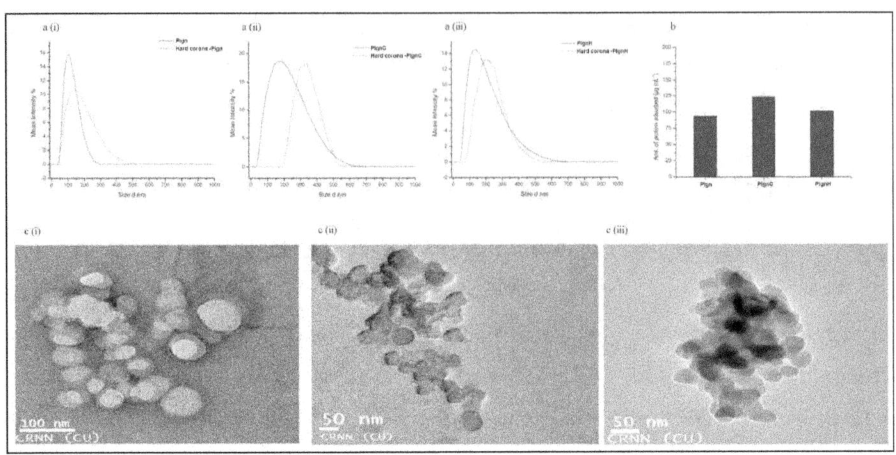

Figure 7.2. (a) Particle size distribution studies for (i) Plgn (ii) PlgnC (iii) PlgnH in absence and presence of human plasma (1 %, v/v) (b) Nanoparticles protein adsorption quantification studies (c) TEM studies of protein hard corona nanoparticles (i) Plgn (ii) PlgnC (iii) PlgnH.

Data recorded from zeta potential analysis (figure 7.3) were also used to understand the dispersion stability of nanoparticles in human plasma. Exposure of biopolymer nanoparticles to human plasma resulted in decrease of charge density upon instantaneous binding of proteins [146]. However, presence of a slight negative charge on the nanosurface at pH 7.4 was likely due to steady sorption of serum albumin [147]. The hard corona forms of PlgnC experienced the maximum electrostatic aggregation effect in plasma media and developed a tendency to precipitate. The results further revealed that least deviation of zeta potential values were observed in case of the Plgn- protein corona, whose stability was facilitated by steric repulsion offered by the amphiphilic nature of the Pluronics stabilized biopolymer [148].

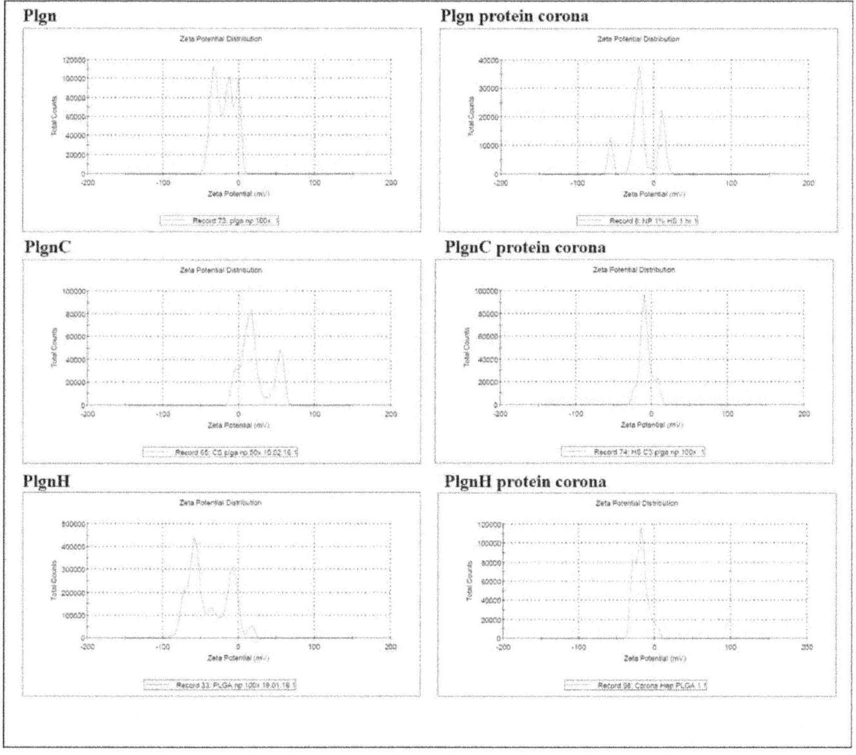

Figure 7.3. Zeta potentials of different nanoparticles and their protein corona forms.

7.3.3. Nanoparticle protein interaction studies

Nanoparticle-protein interactions were also evidenced in fluorescence spectroscopy studies. A steady quenching of fluorescence intensities between 300–450 nm was noted for all three nanoparticle types (figure 7.4). This was attributed to the conformational changes of plasma upon nanosurface adsorption and steric effects due the presence of tryptophan residues in immobilized albumin molecules.

Nanoparticle-protein binding constants in human plasma were obtained from log- log plots based on equation 7.1 for all three types of functionalized nanoparticles.

$$\frac{F_0 - F_S}{F - F_S} = 1 + K_b[nanoparticles] \tag{7.1}$$

Where F_0-F_S, and F-F_S, were the relative fluorescence intensities in absence and presence of nanoparticles. K_b was the nanoparticle-protein affinity constant and the concentration of nanoparticles (µg/ml) was [nanoparticles]. It was found that amongst the different types of nanoparticles, PlgnC exhibited maximum affinity towards human plasma proteins (K_b= 2.96). The binding constants were lesser (K_b= 2.37) in case of PlgnH and minimum (K_b= 1.12) for Plgn. The observations suggested that biopolymeric nanoparticles with different surface chemistry associated differently with the plasma proteins when exposed under the similar conditions.

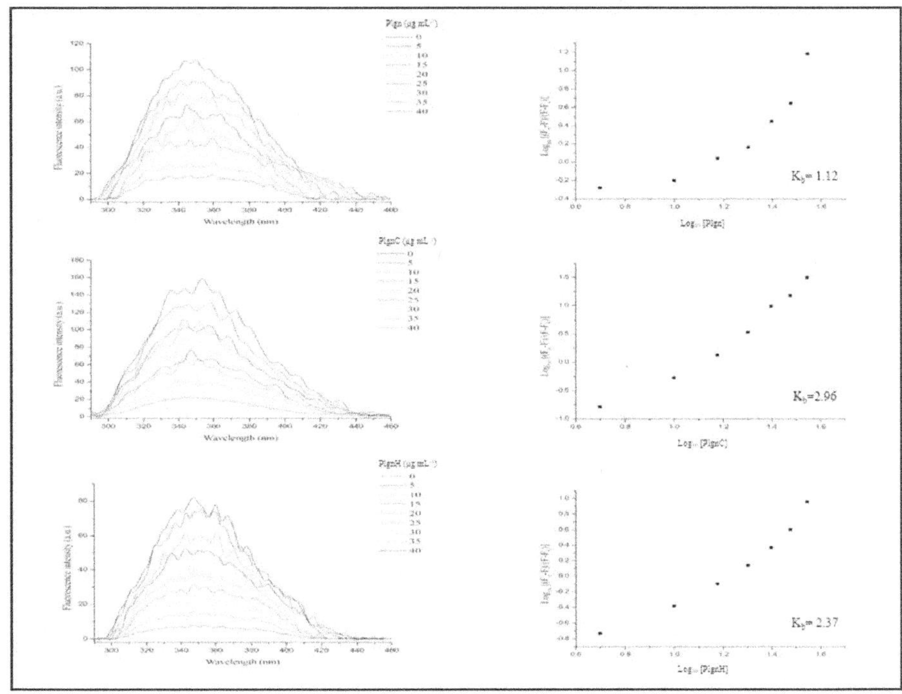

Figure 7.4. (A) Steady state fluorescence quenching of TRP214 signal upon interaction with increasing concentration (0 to 40 µg/ml) of different nanoparticles at 37±0.5 °C (i) Plgn (ii) PlgnC (iii) PlgnH. (B) Binding constant derived from spectral data.

The molecular interactions contributing to the nanoparticle-protein binding were further studied through FT IR spectroscopy studies. The characteristic bands due to amide I at 1630 cm^{-1} - 1658 cm^{-1} (due to C=O stretching), and amide II at 1525 cm^{-1} - 1547 cm^{-1} (due to C-N stretching and N-H in-plane bending) for human plasma protein were noted and monitored for further understanding of nanoparticle-protein interactions. FT IR spectra of Plgn incubated with human plasma presented the protein peaks (for amide I and amide II) at their exact wavenumbers, while preserving the characteristic bands of the nanoparticles (figure 7.5). This suggested that not much protein non-covalent interactions on Plgn surfaces. The corona formation on Plgn was mainly due to adherence during Brownian motion, and proteins self-interactions. In case of the chitosan coated particles, the assigned peaks for chitosan was subdued. However, the plasma protein amide I band shifted from 1650 cm^{-1} to 1644 cm^{-1}, accompanied by amide II band shift from 1538 cm^{-1} to 1551 cm^{-1}. These suggested intermolecular hydrogen bonds formations between the PlgnC carbohydrate groups and plasma protein nanosurfaces [149]. In case, when PlgnH was co-incubated with plasma, there was a marked up-shift (from 1539 cm^{-1} to 1543 cm^{-1}) for the amide II band of protein due to adsorption onto the nanoparticle surfaces. While peaks arising from different groups of Pluronics and PLGA biopolymer (C=O, C-O, C-H) mostly remained intact, a peak loss at 959 cm^{-1} was noted. FT IR results revealed that electrostatic interactions occurred between the negatively charged sulphonates and carboxylates of heparin and amino groups of protein. These protein-glycosaminoglycans interactions required exposure of a binding site in the protein through re-alignment of basic amino acid residues [150] and that might also have contributed in the loss of protein secondary structures.

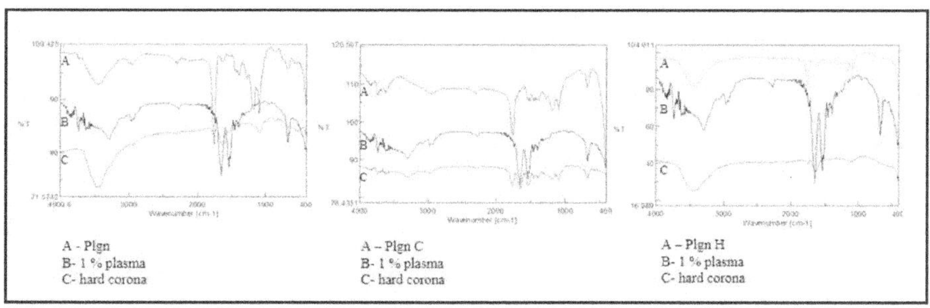

Figure 7.5. FT IR studies of different nanoparticle types and hard corona.

7.3.4. Analysis of nanoparticle-bound proteins

Nanoparticle-protein binding constants and underlying chemical interactions at the nano-bio-interfaces ostensibly are the major factors which determined the number and types of proteins comprising the hard corona structures [151]. Proteins present in the corona of the nanoparticles were hence studied, through gel electrophoresis (figure 7.6a). This was considered to comprehend the nature of protein deposition on differentially functionalized nanoparticles. Coomassise stained gels for human plasma showed typical bands arising out of the most abundant proteins such as transferrin (80 kDa) [152], albumin (67 kDa), IgG (heavy chain, 50 kDa), apolipoprotein E-1 (34 kDa) [147,153]. Electrogram arising from the surface functionalized nanoparticles coronas revealed that the extent of adsorption of all major plasma proteins differed for different nanoparticle type.

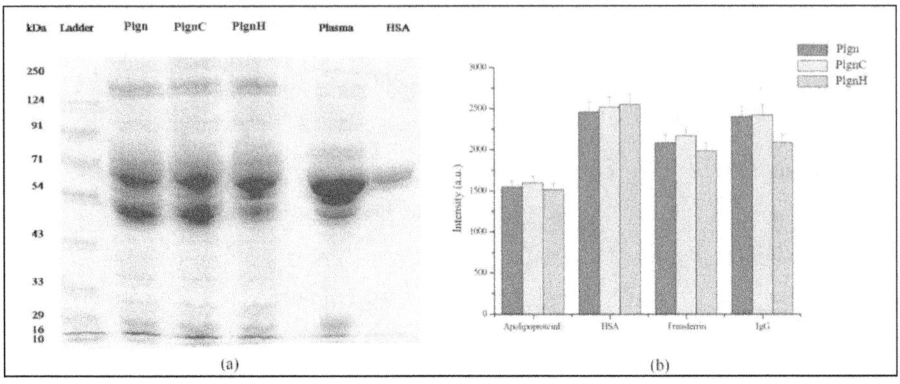

Figure 7.6. Analysis of human plasma protein on nanoparticles (a) Resolved SDS-PAGE of plasma proteins adsorbed onto nanoparticle types (b) Band intensity histograms for adsorbed major proteins Data were represented as mean±S.D. (n=3).

The SDS-PAGE studies confirmed variable band intensity for proteins on nanoparticle types (figure 7.6b). Serum albumin being the most abundant plasma protein (~44 g/l), was adsorbed onto the all nanoparticle types. Albumin attachment to PlgnH was comparatively more as a result of its high surface charge (-38.4 mV). Other proteins such as IgG, transferrin and apolipoprotein E-1 were strongly attracted towards the heparin coated nanosurface. The relative quantities of these three proteins were higher on chitosan coated nanosurface. Chitosan bears positively charges in low pH, and can bring out innate immunological responses within biological systems [154]. Hence, PlgnC were likely to associate more with immunoglobulins (IgG) than the other nanoparticles. This was considerably lower in case of PlgnH because of heparin attachment. Heparin on nanoparticles surfaces ensured reduced activation of the complement system [155]. Protein adsorption on Plgn surface was minimum, as previously recorded through the Bradford assay method (figure 7.2b). These observations rationalized the results obtained from the detailed plasma protein-nanoparticle interaction studies through FT IR and fluorescence spectroscopy. The hard corona surrounding the nanoparticles was composed of proteins of different types and amounts, the binding of which depended mostly upon the biopolymer nanosurface chemistry.

Freshly reconstituted human plasma demonstrated but a low ThT fluorescence response (figure 7.7A). This was due to the presence of proteins, such as microglobulins and glycoproteins possing in β sheet conformations [156,157]. Plasma incubated at 60±2 °C for 1 hour caused a large increase in ThT fluorescence intensity at 480 nm. Albumin experiences conformation changes to form amyloid-like structures when incubated at high temperatures [158], and that was mainly responsible for the increase of ThT fluorescence response. This observation was considered as positive control. Human plasma co-incubated with Plgn beared a low ThT intensity response which indicated that a fewer number and mass of proteins that were adsorbed on the nanosurface faced minimal conformational changes [159]. Among the differently functionalized nanoparticles incubated with plasma, ThT response for PlgnC was recorded as maximum. Chitosan on nanoparticles surfaces influenced protein amylodogenic β structure formations. This observation matched well with the PlgnCs huge increase in hydrodynamic diameter (7.3.2., 171.2 %) due to corona formations. Similar observations were reported during the interaction between chitosan and albumin [160]. In a way, chitosan functionalized nanoparticles are likely to influence an amyloidogenic protein reactions and induce adverse effects on longer term applications. Chitosan functionalization led to β sheet formations of the corona proteins in the vicinity. A certain degree of cross β sheet structures were also detected upon association of PlgnH with plasma proteins. This was consistent with the fact that the heparin molecules facilitate protein fibrillation [161,162]. It was thus convincing that the nanoparticles corona formation known in plasma contact is possibly induced by β sheet deposition, nucleation and fibrillation. Protein corona around the PlgnC and PlgnH comprised mostly of similarly aggregated proteins. It appeared that Plgn demonstrated minimal effects on β sheet deposition and protein fibrillation.

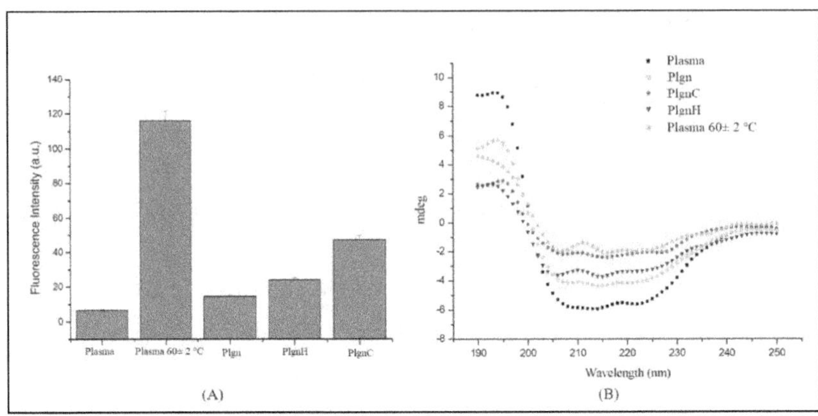

Figure 7.7. (A) ThT fluorescence studies on 1 % v/v blood plasma co-incubated with and without nanoparticles. Intensities expressed as mean±S.D. (n=3). (B) Far UV CD spectra of 1 % v/v blood plasma co-incubated with or without nanoparticles.

CD studies were also applied for a better understanding and to measure the secondary structural transitions occuring in plasma proteins upon interactions with the surface modified nanoparticles. Negative peaks at 208 and 222 nm in the spectrum of Plgn-protein corona aroused due to $\pi \to \pi^*$ and $n \to \pi^*$ transitions of carbonyl groups present in the polypeptide chains of plasma protein albumin [121]. Plasma incubated at 60±2 °C exhibited minimization of the negativity peaks at 208 and 222 nm indicated disruption of α helix contents (figure 7.7B). Presence of a single negative peak at 215 nm was due to increase of β sheet forms [85,119]. Plasma co-incubation with Plgn exhibited minimum changes in the CD spectrum in comparison to fresh plasma. This corroborated in line with other studies (section 7.3.2 and 7.3.3). Conformational changes in proteins therefore were unlikely due to the absence of nucleation inducers. Considerable deviations in terms of mdeg measured for PlgnC and PlgnH were recorded. Both PlgnC and PlgnH perturbed the native conformations of protein present in the vicinity of the nanoparticles. The ThT fluorescence studies and the CD spectroscopy thus demonstrated that the nanosurface chemistry influences native protein conformations in the corona microenvironment.

7.4. Summary

This part of the work embarked on understanding influences of biopolymer nanoparticle surface chemistry on the plasma proteins in the vicinity. PLGA biopolymer was used as the base and non-covalent functionalization that are raft in drug delivery devices design were taken in consideration. Implications of biopolymer nano-chemistry appeared enormous in case of plasma protein interactions, hard corona formations and induced protein fibrillation.

The thickness and composition of the hard protein corona varied significantly with the biopolymer nanoparticles surface chemistry. It was further observed that the protein adsorption on nanosurface altered the native protein structure, and led to formation of amyloid-like aggregates and hard corona formations in most cases. Pluronics surface stabilization for PLGA greatly restrained protein fibrillation and corona formation around nanoparticles. This part of the work adressed some of the contagious issues, such as the biopolymer nanoparticles behaviour in plasma protein interfaces, effects of surface chemistry on hard corona formations and influences of biopolymer nanochemistry on protein fibrillations. The work also presented some suggestive strategies and biopolymer nanoparticles chemistry requirements for unhindered transport in plasma environment.

CHAPTER 8

Biopolymer Nanoparticles against Drug Induced Hepatotoxic Conditions

8.1. Introduction

Accumulation of misfolded proteins in cells and the tissue sites often inflict irreversible damages, leading to cell injury and tissue toxicity. Impaired metabolism of amyloidogenic proteins further contribute in activation of various signalling and infuse stress damages [163]. These also bring upon apoptosis and demonstrate tissue specific toxic reactions [164,165]. Liver is a highly perfused organ and is the first line of defence against external stimuli such as toxins, alcohol and xenobiotics [166,167]. Hepatocyte endoplasmic reticulum (ER) is one of most active sites responsible for folding of synthesized proteins and transport. ER stress and injury to the hepatic tissues are hallmarks of dilapidating diseases such as cystic fibrosis, drugs like paracetamol induced liver injury, and alcohol liver damages [168,169]. Paracetamol overdose in rodents thus extended a readily adoptable model for understanding the biochemical effects of drug induced oxidative stress, cell damages and liver injury [170].

Inflammation of the liver and associating amyloidogenic fibrillar deposition is one fundamental pathophysiology in case of a number of hepatotoxic conditions. Progressive liver damage due to various drug induced liver diseases are increasingly being linked to site specific amyloid proteins fibrillation. In fact, the term 'amyloid' was first coined by Rudolf Virchow in 1854 while performing autopsy of deceased who suffered from chronic hepatic inflammation [3,4]. Iodine strained starchy (Latin: *amylum)* deposits originally observed, were later turned out to be nitrogen rich β structured protein fibrils resistant to proteolysis.

Quercetin (QR) demonstrated high fibril inhibitory activity. Besides, QR exhibited a degree of disruptive effects on the mature fibrils. QR and sustained release nanoparticle carriers were hence considered in this part of the work for therapeutic interventions. QR suffers from biopharmaceutical limitations for lack of solubility and also encounters stability issues in physiological environment [171]. An approach for QR solubility enhancement by complexation with cyclodextrin was studied and that reportedly caused nephrotoxicity [172]. Nanoparticulation of QR has therefore been perceived as an alternative to enhance the systemic availability and deliver bioactives to molecular targets due to enhanced permeation and retention (EPR) effects [144]. It was envisaged that quercetin in nano-carriers would be efficacious in mitigation drug induced complications of liver in physiological conditions.

This work part was meant to design and synthesize biopolymer nanoparticles for slow release of QR cargo. QR nanoparticle cargo devices were further evaluated in drug (paracetamol) induced liver injury in mice model. Biopolymer nanoparticles loaded with QR appeared as one successful agent in both *in vitro* and *in vivo* studies.

8.2. Experimental

8.2.1. Preparation of nanoparticles and characterization

QR loaded PLGA nanoparticles (QPlgn) and PLA nanoparticles (QPln) were prepared following oil-in-water emulsion and evaporation method [173]. In a generalized protocol, 10 mg of biopolymer and 2 mg of QR were dissolved together in 1 ml of methylene chloride (HPLC grade, Spectrochem, India). The organic phase was rapidly poured into 10 ml of 1 % w/v aqueous Pluronic F-127 solution and the mixture was sonicated for 2 minutes in a 700 MW probe sonicator (Vibra cell VCX 750, Sonics, USA). The resultant o/w emulsion was diluted to 20 ml with water and homogenized (TH 02, Omni International, USA) further for 20 minutes at 20,000 rpm under external cooling in ice water. The solvent evaporation was continued for 12 hours more over a magnetic stirrer at room temperature. The nanoparticles formed were harvested by centrifugation at 16,000 rpm for 60 minutes at $4\pm2°C$, washed with water, re-centrifuged, and preserved in desiccators. Bare nanoparticles (Pln and Plgn) devoid only of QR were also prepared similarly.

The hydrodynamic diameter and polydispersity index (PDI) of nanoparticles were determined in a zetasizer (section 6.2.1). Batch measurements were performed in triplicates at 25 ± 2 °C and the average was recorded. The particle electrophoretic mobility was also determined against an applied electrical field and data averaged for analysis. Nanoparticles size studies were conducted in transmission electron microscopy as well. Diluted samples were placed on carbon coated copper grids, negatively stained with uranyl acetate solution, air dried and TEM images captured for analysis.

8.2.2. Quercetin analysis

QR was analysed in a reverse phase HPLC system fitted with Waters 515 pumps (Waters, USA), ODS column 250 x 4.6 mm (Phenomenex, USA), and 2996 PDA detector (Waters, USA) set at 255 nm. Isocratic solvent flow at 1 ml/min was used using acetonitrile: water (40: 60, v/v). A peak area (y) versus concentration (x) graph, y= 40367x - 38914, R^2=0.998, was first developed from standard QR injections and that was used for estimation throughout.

Nanoparticle cargo loading was determined from the differences in QR analysis in solutions, before and after nanoparticulation. Percentage encapsulation efficacy (% EE, equation 8.1) was derived from the results of three batch experiments.

$$\% \text{ EE} = \frac{\text{Mass of QR originally taken - Mass of QR in supernatant}}{\text{Mass of QR originally taken}} \times 100 \tag{8.1}$$

QR release from the both QPlgn and QPln over time was studied in separate setups at 37±0.5 °C using dialysis bags, molecular weight cut-off 12,400 Da (Sigma, USA). QR loaded nanoparticles equivalent to about 5 mg of payload dispersed in 1 ml of PBS and transferred into end-tied dialysis bags. Each bag was placed in glass vials containing 35 ml of PBS and the glass vials were placed on an orbital shaker set at 50 rpm, 37±0.5 °C. Aliquots of 5 ml were withdrawn from the vials at pre-determined time intervals and were replaced with the same amount of fresh medium to maintain sink conditions. QR released at each time points were quantified from 20 µl aliquot injections into the HPLC set up. The standard curve equation was used for QR quantification with necessary corrections for dilution factors. The percentage cumulative mass (y) release over time (x) was derived from three batches of experiments and the data were averaged for x-y plots. Korsmeyer-Peppas model was used for understanding the molecular release parameters.

8.2.3. Hepatotoxicity studies

8.2.3.1. Animals

Male Swiss albino mice (30-35 g) were housed in polypropylene cages and were allowed to acclimate to laboratory conditions (25±2 °C, 65±5% relative humidity and 12 hours light/ dark cycle) for 15 days prior to *in vivo* experiments. They were fed with standard pellet diet (Hindustan Unilever, Mumbai) twice daily and water *ad libitum*. Only water was permitted on the day of the experiment. The animal experiments were conducted in accordance with guidelines of National Institutes of Health guidelines for care and use of laboratory animals (NIH Publications No. 8023, revised 1978) and the studies were sanctioned by the institutional animal ethical committee of Dr. B. C. Roy College of Pharmacy and Allied Health Sciences, Durgapur (BCRCP/IAEC/13/2018).

8.2.3.2. Hepatotoxicity: induction and evaluations

Animals were weighed and assigned randomly to seven groups of six mice each. The group I received saline only for 7 days, group II was administered with a single intra-peritoneal (i.p.) dose of paracetamol (300 mg/kg body weight). Group III was treated i.p for 7 days with 20 mg/kg body weight of QR dispersed in 0.1 % v/v of aqueous Tween 80 prior to paracetamol (300 mg/kg) administration i.p. [174,175]. Group IV received bare nanoparticles Pln, dispersed in 0.1 % v/v Tween 80 for 7 days prior to paracetamol (300 mg/kg) i.p. Group V received QPln equivalent of 20 mg/kg QR payload dispersed in 0.1 % v/v Tween 80 for 7 days, prior to paracetamol (300 mg/kg) i.p. Group VI received bare nanoparticles Plgn dispersed in 0.1 % v/v Tween 80 for 7 days, prior to paracetamol (300 mg/kg) i.p. and Group VII received QPlgn equivalent of 20 mg/kg QR payload dispersed in 0.1 % v/v Tween 80 for 7 days, prior to paracetamol (300 mg/kg) i.p. Animals were euthanized 24 hours after paracetamol administration, blood sampled and liver part collected for analysis.

Blood samples were processed and serum was collected after centrifugation at 3000 rpm, for 5 minutes at 4±2 °C. Serum markers such as serum glutamic pyruvic transaminase (SGPT), serum glutamic oxaloacetatic transaminase (SGOT), alkaline phosphatase (ALP), bilirubin contents and cholesterol levels were quantified using specific reagent kits (Ecoline, Merck Specialities, India). Liver sections were collected, thoroughly washed, dipped in 10 % v/v formalin and embedded in paraffin blocks for microtomy. Tissue samples were further stained with haematoxylin and eosin for microscopic analysis.

8.2.4. Statistical analysis

All experiments were performed in triplicates and the data were presented as mean ±S.D. Student's t-test was performed for comparative analyses of mean values. Differences were considered significant when $P<0.05$.

8.3. Results and discussion

8.3.1. Synthesis and characterization of nanoparticles

Nanoparticles were synthesized following emulsion and evaporation technique. Various trial and error runs were performed and emulsion evaporation technique was preferred over solvent diffusion technique for higher loading of QR bioactive cargo. Methylene chloride is a common solvent for PLGA, PLA and QR [176]. Pluronic F-127 was preferred as stabilizer molecule for experimental ease and due to its null effects on protein precipitation and nanoparticle corona formation [177]. Cargo loaded PLGA biopolymer produced nanoparticles, having an average hydrodynamic diameter of 180±12.76 nm and a low PDI of 0.294. Nanoparticles from PLA had slightly larger hydrodynamic diameter of 225±16.23 nm and the PDI was 0.359 (figure 8.1A). The hydrodynamic diameters of bare nanoparticles Pln and Plgn were recorded as 160±10.62 and 125±9.45 nm. This was likely due to no QR loading. The zeta potentials for QPlgn and QPln were observed to be -31.2± 1.33 and -25.5±1.28 mV, respectively. The negative surface charges on nanoparticles were understandably due to Pluronic stabilization and the exposed carboxylic groups of the biopolymers. Electron microscopic studies revealed that the QR incorporated nanoparticles (both QPlgn and QPln) prepared through emulsion evaporation-homogeneous technique were in uniform size (figure 8.1B). This appeared consistent with similar other studies [176].

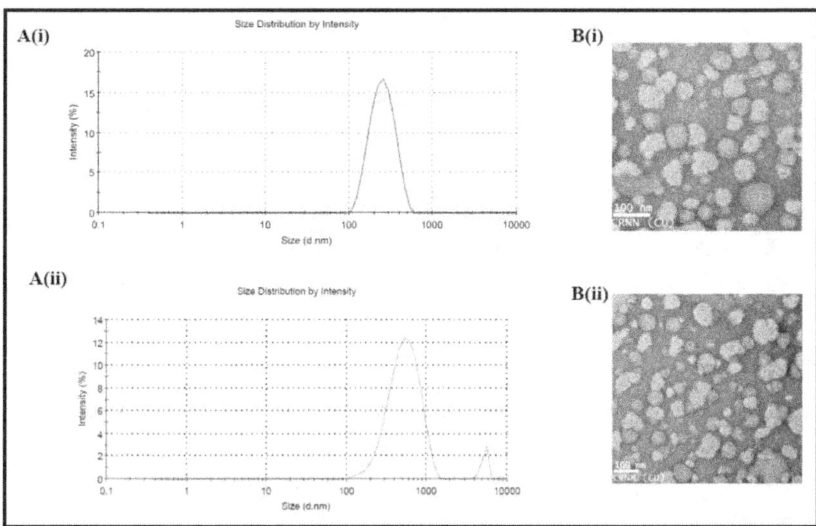

Figure 8.1. Characterization of (i) QPlgn and (ii) QPln. (A) hydrodynamic diameter (nm) (B) electron micrograph.

8.3.2. Quercetin release and entrapment efficiency

QR was quantified using a reverse phase HPLC setup. Elution time of QR was 6.8 minutes and a standard equation developed from the response of different concentration injections were used for quantifications. Supernatants collected before and after cargo loaded nanoparticles harvesting, were injected in HPLC. QR entrapment efficiency (% EE) for QPlgn and in QPln were 94.66±0.39% and 97.4±0.41%, respectively. This indicated that emulsion evaporation method prevented in-process loss of cargo molecules. Relative lipophilicity of PLA has likely contributed in enhanced drug loading in case of QPln.

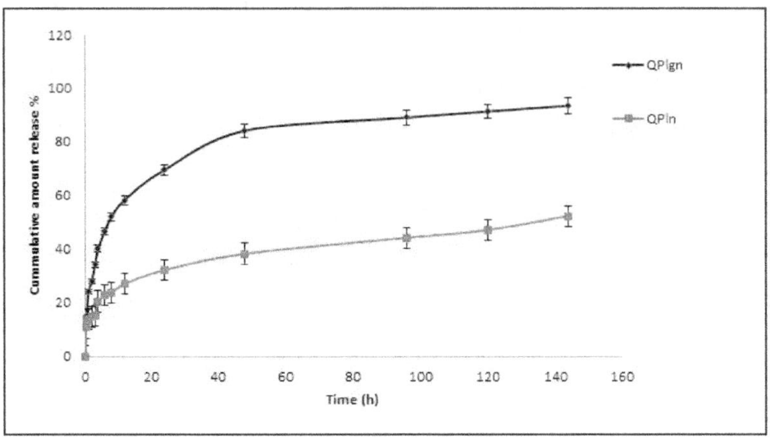

Figure 8.2. QR release from QPlgn and QPln at 37±0.5 °C *in vitro*. Cumulative release percentages expressed as mean±S.D. (n=3).

QR release from nanoparticle types (figure 8.2) were biphasic in nature, comprising of an initial burst phase followed by a near steady state stage. The initial burst effect was recorded near 4 hours in both cases. This was likely due to nanoparticle surface adherent molecules. QR cumulative release from QPln and QPlgn was recorded as 52±2.73 % and 93±3.14 % over a period of 144 hours. QPln represented a sustained delivery but only 52 % of QR load can be recorded during the *in vitro* release study period. Cargo mass loading in QPln was however in higher ranges compared to QPlgn. Korsmeyer-Peppas model (8.2) was applied in order to understand the payload release mechanism from the nanoparticle delivery systems.

$$M_t/M\alpha = Kt^n \tag{8.2}$$

Here, Mt was the QR cargo released at time 't', and Mα was the total QR payload. Also, 'n' was the release exponent, describing fundamental mechanisms and K represented the release rate constant in time^{-1}. The n and k values recorded were tabulated in table 8.1. In near spherical nano-matrices when n<0.4, a Fick's mass diffusion mechanism for QR was considered prevalent. The release rate for QPln was considerably slower when compared to that of QPlgn. Polymer relaxation under *in vitro* conditions was apparently very minimal. QR has extremely low water solubility (0.00215 mg/mL to 0.00263 mg/mL for dehydrate) at 25 °C, and that was considered as one reason for QR sustained release under *in vitro* conditions [178].

Table 8.1: Release mechanism for quercetin loaded nanoparticles

Quercetin loaded nanoparticles	Korsmeyer-Peppas model parameters		
	R^2	n value	K value
QPlgn	0.9619	0.3169	0.31
QPln	0.9743	0.2523	0.1395

8.3.3. In vivo studies in paracetamol induced hepatotoxicity

QR cargo nanoparticles were evaluated using paracetamol induced mouse hepatotoxicity model. Liver is a major metabolic organ. Prognosis and diagnosis of hepatotoxic conditions is complex and are still very hypothetical [179,180]. Paracetamol extend a definitive inflammatory model pertaining to the liver tissue. The drug is metabolized in the liver by the oxidase enzymes, CYP2E1 to form NAPQI (N-acetyl-p-benzoquinoneimine) [181,182]. NAPQI is highly reactive and when present in excess, it readily reacts with liver proteins and serum albumin. This induces oxidative stress and is diagnosed as paracetamol hepatotoxicity [183,184]. Monitoring protein paracetamol metabolite adduct is currently explored as one useful technique for better management of overdose hepatotoxicity.

Albumin is predominantly bio-synthesized in liver [185]. Serum albumin is a charged macromolecule and that serves a multitude of functions including acid-base homeostasis, anti-inflammatory, drug binding (favourable or unfavourable) and osmotic activity due to Gibbs–Donnan effect [186]. Albumin has a distinctive relationship particularly in case of decompensated liver diseases. It has been shown earlier that variation in serum albumin is linked distinctly with drug induced hepatotoxicity, like that in case of paracetamol injury of liver tissues [187,188]. Current prognosis for drug induced liver damages due to Amatoxin mushroom poisoning, industrial solvents, paracetamol and isoniazid alike are strongly linked to serum albumin levels and deformities. Interestingly, Virchow (1854) originally demonstrated amyloidosis in liver autopsy [3,4]. Now, new drugs are increasingly being explored to target various stages in the pathogenesis of different types of fibrils in liver, serum, brain tissues and others, thus raising new hope for patients with obstinate conditions.

Table 8.2: Effects of QR and nanoparticles against paracetamol induced hepatotoxicity

Group	SGOT (IU/L)	SGPT (U/L)	ALP (U/L)	Total Bilirubin (mg/dl)	Mortality
I	26.1± 2.1	31.1± 2.5	157.7± 9.3	0.74± 0.06	0/6
II	320.1± 21.3	289.3± 19.8	350± 24.3	1.85± 0.1	2/6
III	181.1± 12.5*	156.1± 8.9*	195± 13.7*	1.05± 0.09	0/6
IV	300.3± 20.6	262.5± 16.6	358.2± 21.8	1.81± 0.09	2/6
V	39.± 3.8	51.1± 3.6	169.3± 9.1	0.87± 0.07	0/6
VI	311.4± 22.3	271.2± 11.6	350.1± 21.4	1.82± 0.08	2/6
VII	28.6± 3.1	33.1± 3.4	151.1± 8.6	0.83± 0.07	0/6

Notes: Results expressed as mean±S.D. (n=6); *$P<0.05$ compared to paracetamol treated group.

In order to estimate the degree of injury in test groups, different serum marker levels were analysed. A single dose of 300 mg/kg paracetamol i.p., caused massive increase of serum glutamic pyruvic transaminase (SGPT), serum glutamic oxaloacetic transaminase (SGOT) and alkaline phosphatase (ALP) levels (table 8.2) in group II in comparison to the normal control (group I). Rise of serum SGOT and ALP levels were due to loss of hepatic structural integrity, cellular permeability changes and increase of aminotransferase synthesis [189]. Increase in the serum bilirubin levels also indicated disturbances in normal liver functions, leading to cholestasis [190]. QR pre-treatment in group III protected the liver against paracetamol induced injury which was apparent from the reduced serum markers levels ($P<0.05$). This observation was in agreement with some previous studies on QR in liver injury models [191,192]. Nanoparticles without QR payload (group IV and VI) did not cause any change in serum marker levels. Serum marker levels were but much reduced and were towards normal upon pre-treatment with QPln (group V). Pre-treatment with QPlgn (group VII) also restricted the release of hepatic enzymes and maintained the ALP, SGOT, SGPT at lower ranges. Perhaps hepatoprotection was prevalent due to sustained exposure of QR from the nanoparticles present in the hepatic tissues. QR treatment, on the other hand, failed to provide maximum protection likely due to QR transport limitations and low distribution half-life [193].

Histopathological data were also obtained to substantiate efficacy evaluations. Paracetamol overdose resulted in severe damages in centrilobular region, caused dilation of sinusoids, disintegrated cell membranes and evidenced appearance of inflammatory cells (figure 8.3B). Marked hepatic injury was recorded [194]. Animals treated with bare nanoparticles showed similar results (figure 8.3C and E). QR however extended hepatoprotective activity akin to that in earlier reports [195,196]. Significant number of hepatocytes were present and the central vein appeared distinct in case of both QPln and QPlgn treated groups (figure 8.3D and F). Incidence of lesions or presence of monocytes were not visible in hepatic architecture. The ballooning of hepatocytes was also restricted. These results showed that quercetin in nano forms more effectively protected the native hepatocellular structures from injury in comparison to the crude QR.

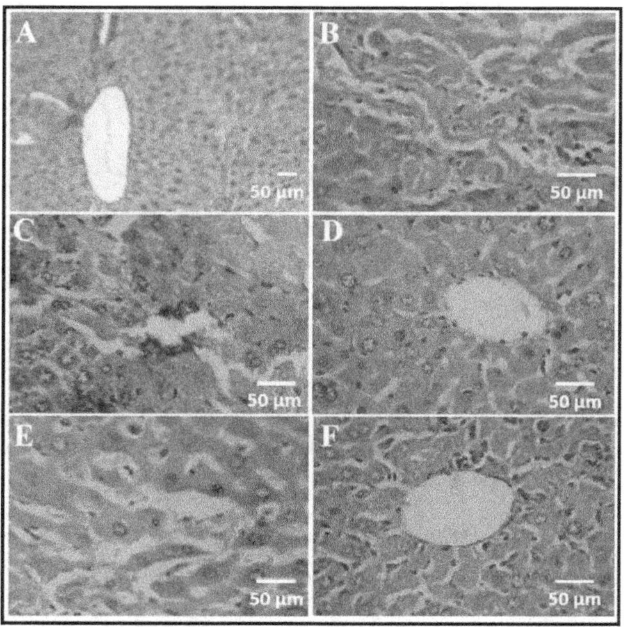

Figure 8.3: Mouse liver histology (x10) after hematoxylin-eosin staining. (A) Normal control, (B) paracetamol treated, (C) bare Pln + paracetamol, (D) QPln + paracetamol, (E) bare Plgn + paracetamol, (F) QPlgn + paracetamol. Scale bar 50 µm.

Studies have revealed earlier that paracetamol overdose cause leakages of electrons in the mitochondria electron transport chain and form misfolded protein adducts, leading to generation of free radicals and membrane damage [170]. Various preventive agents and paracetamol derivatives are currently under intense studies as palliatives and for treatment of various decompensated liver diseases [197]. QR is one of the best known plant derived antioxidants and can inhibit protein aggregates and disrupt misfolded protein adducts. Sustained delivery of QR in mouse models indicated protection of liver tissues against oxidative stress induced damages.

8.4. Summary

Engineered nanoparticles with QR cargo were developed. A facile emulsion solvent evaporation technique was adopted for synthesis of cargo loaded PLGA and PLA matrixed nanoparticles. Pluronic F-127 was established as one very useful nanoparticle stabilizer. Solubility limitation of QR was overcame and sustained release nanoparticles appeared useful against drug induced hepatotoxic conditions. Engineered nanoparticles restored serum hepatic markers to near normal levels.

CHAPTER 9

Conclusion and Perspectives

9. Conclusion and Perspectives

The chaotic motions of proteins in bio-systems present some of the hallmark events both in favour of functionality and toxicity [198]. Proteins structural plasticity and conformational flexibility permit them often to perform incredible tasks in biological environment. A spectrum of binding scenarios is nonetheless passable in similar performing proteins [199]. Understanding the relationships between amino acid sequence, protein structure and functionality represent one of the major challenges today in protein sciences. This area is also currently in focus for finding some solutions in various dilapidating disease conditions. These are termed together as amyloidogenic diseases. Amyloids were originally ascribed to some iodine strained waxy deposits found in inflamed human liver [3,4]. In general, these are ordered aggregates of diameter 100–200 Å, and are comprised of arrays of intermolecular β sheet structures running parallel to a long axis of protein fibrils. Various functional proteins such as insulin, serum albumin or α synuclein present similar traits. Standardized techniques are presently available to aid in recognizing amyloids by staining with specific dyes, such as thioflavin T and other derivatives. The mechanism of protein misfolding and aggregation is largely understood by a seeding–nucleation model [200]. During this, a thermodynamically unfavourable slow nucleation phase is followed by a fast elongation stage for mature protein fibrils. The phenomenon is graphic and could be easily recorded, both, for the fundamental understandings *in vitro* and subsequent molecular interventions.

This work entails a pursuit for exploring in protein fibrillation intricacies, simultaneously with a search for some plant bioactive molecules and engineered nano-carriers for efficient interruption of protein fibrillation. Human serum albumin, a functional protein in blood was used as a protein model for getting to a depth of understanding at various levels. Both the wet lab techniques and *in silico* studies were performed in most cases.

AG (andrographolide) from *Andrographis paniculata* demonstrated concentration-dependant inhibition of HSA fibrillation. The inhibitory effects appeared principally due to favourable and perhaps reversible hydrogen bonding as well as site specific hydrophobic interactions with serum albumin in its flexible forms. Similar labdane diterpenoids were reported for concentric anti-inflammatory activities [201]. The observation therefore opened vistas for similar molecular developments.

QR (quercetin), a flavonol, is widely promulgated for alleviating in diabetes, Alzheimer's disease and various other life style disorders. Effects of QR on HSA fibrillation appeared remarkable. QR inhibited HSA fibrillation to an extent of ~44 %. The molecule extended multiple hydrogen bonds, which in turn has stabilized the native protein conformation in solutions. Uniquely, QR also facilitated disruption of preformed and mature fibrils in a concentration-dependant manner. QR remained a molecule under intense studies. Inhibitory potentials in case cardiac and/or renal fibrosis were revelling [202,203]. Systemic delivery of QR has though remained a contagious issue. This study extended some insight into QR diffusion in protein interfaces so much so as to lessen fibrillation and possibly effect in detoxification of mature fibrils.

SB (silybin), is a flavonolignan obtained from the milk thistle plant. The molecule is a therapeutic of choice in acute liver damages. SB repressed HSA fibrillation, and the maximum inhibitory effect was observed when HSA:SB ratio was at 1:1.15. Binding of SB through hydrogen bonding as well as hydrophobic interactions was responsible for stabilization of native protein conformations, delaying formation of nuclei and inhibiting the oligomeric intermediates formation.

Gold nanoparticles *per se* are efficient tools for directed interactions in protein interfaces. One pot synthesis for size controlled gold nanoparticles was achieved under sonication and no toxic surfactant/stabilizers were used. Polyphenol stabilized gold nanoparticles, GAuNPs, QAuNPs and SAuNPs were fully characterized and were used against protein fibrillation. Gold nanoparticles effects on HSA fibrillation were clearly due to rapid diffusion, surface functionalization and Brownian motion. Gold nanoparticles stabilized in borate or citrates were explored in α-synuclein aggregation [204]. These molecules have opposing effects in protein folding-misfolding interfaces. Polyphenol capped gold nanoparticles but extended a newer dimension for biosafe gold nanoparticles applications as fibrillation inhibitors.

Nanoparticles often withered in bio-systems due to protein deposit corona formations. Various proteins misfolding during diffusive transport of nanoparticles in plasma environments were envisaged as a driving reason. Modulating biopolymer nanoparticles surface chemistry and charge were considered as a way to take control. Pluronics as a stabilizer on PLGA nanoparticles was effective in restraining hard corona formation and inhibited protein fibrillation as well. Nanochemistry at the carrier surfaces may be applied to thwarted protein adsorption and slacken corona formations.

Biopolymer nanoparticles with bioactive QR cargo were devised for some *in vivo* studies. Pluronics was used as a stabilizer throughout. QR is a low soluble, low permeable compound, and this has affected drug-like applications of QR throughout [205]. Sustained release QR cargo nanoparticles were tested against paracetamol overdose hepatotoxic conditions. Designer QR nanoparticles protected the liver tissues against drug induced liver injury and conserved the hepatic biomarkers levels to near normal.

Fibrillating proteins are currently unveiling as one fundamental cause in a range of degenerative diseases. Amyloid protein aggregations are related to nearly twenty (20) diseases and that list is expanding as well. Various oligomeric intermediates and supra-molecular assemblies are considered for toxic insoluble fibrils formations in similar disease conditions. This work has explored the effects of some plant bioactives and nano-carriers on protein fibrillation. In depth studies have discovered successful application of polyphenolics such as QR, AG and SB for inhibition and disruption of protein fibrils. Pluronics was established as a biopolymer nanoparticle stabilizer useful for hard corona retardation, fibrillation inhibition and payload delivery. Gold nanoparticles stabilized in polyphenolics also unravelled another strategy for intervention in fibrillating protein.

References

References

[1]   P. Bharmoria, D. Mondal, M.M. Pereira, M.C. Neves, M.R. Almeida, M.C. Gomes, J.F. Mano, I. Bdikin, R.A.S. Ferreira, J.A.P. Coutinho, M.G. Freire, Instantaneous fibrillation of egg white proteome with ionic liquid and macromolecular crowding, Commun. Mater. 1 (2020) 34. doi:10.1038/s43246-020-0035-0.

[2]   J. Tyedmers, A. Mogk, B. Bukau, Cellular strategies for controlling protein aggregation, Nat. Rev. Mol. Cell Biol. 11 (2010) 777–788. doi:10.1038/nrm2993.

[3]   R.N. Rambaran, L.C. Serpell, Amyloid fibrils, Prion. 2 (2008) 112–117. doi:10.4161/pri.2.3.7488.

[4]   R. Kisilevsky, S. Raimondi, V. Bellotti, Historical and Current Concepts of Fibrillogenesis and In vivo Amyloidogenesis: Implications of Amyloid Tissue Targeting, Front. Mol. Biosci. 3 (2016) 17. doi:10.3389/fmolb.2016.00017.

[5]   Y.-C. Shen, C.-F. Chen, W.-F. Chiou, Andrographolide prevents oxygen radical production by human neutrophils: possible mechanism(s) involved in its anti-inflammatory effect, Br. J. Pharmacol. 135 (2002) 399–406. doi:10.1038/sj.bjp.0704493.

[6]   F. Li, E.M. Lee, X. Sun, D. Wang, H. Tang, G.-C. Zhou, Design, synthesis and discovery of andrographolide derivatives against Zika virus infection, Eur. J. Med. Chem. 187 (2020) 111925. doi:10.1016/j.ejmech.2019.111925.

[7]   P. Roy, S. Das, A. Mondal, U. Chatterji, A. Mukherjee, Nanoparticle Engineering Enhances Anticancer Efficacy of Andrographolide in MCF-7 Cells and Mice Bearing EAC, Curr. Pharm. Biotechnol. 13 (2012) 2669–2681. doi:10.2174/138920112804724855.

[8]   X. Kang, Z. Zheng, Z. Liu, H. Wang, Y. Zhao, W. Zhang, M. Shi, Y. He, Y. Cao, Q. Xu, C. Peng, Y. Huang, Liposomal Codelivery of Doxorubicin and Andrographolide Inhibits Breast Cancer Growth and Metastasis, Mol. Pharm. 15 (2018) 1618–1626. doi:10.1021/acs.molpharmaceut.7b01164.

[9]   Y.-H. Hsu, Y.-L. Hsu, S.-H. Liu, H.-C. Liao, P.-X. Lee, C.-H. Lin, L.-C. Lo, S.-L. Fu, Development of a Bifunctional Andrographolide-Based Chemical Probe for Pharmacological Study, PLoS One. 11 (2016) e0152770.

doi:10.1371/journal.pone.0152770.

[10]   V.S. Nguyen, X.Y. Loh, H. Wijaya, J. Wang, Q. Lin, Y. Lam, W.-S.F. Wong, Y.K. Mok, Specificity and Inhibitory Mechanism of Andrographolide and Its Analogues as Antiasthma Agents on NF-κB p50, J. Nat. Prod. 78 (2015) 208–217. doi:10.1021/np5007179.

[11]   J.Y. Seo, E. Pyo, J.-P. An, J. Kim, S.H. Sung, W.K. Oh, Andrographolide Activates Keap1/Nrf2/ARE/HO-1 Pathway in HT22 Cells and Suppresses Microglial Activation by Aβ(42) through Nrf2-Related Inflammatory Response, Mediators Inflamm. 2017 (2017) 5906189. doi:10.1155/2017/5906189.

[12]   R. Yang, S. Liu, J. Zhou, S. Bu, J. Zhang, Andrographolide attenuates microglia-mediated Aβ neurotoxicity partially through inhibiting NF-κB and JNK MAPK signaling pathway, Immunopharmacol. Immunotoxicol. 39 (2017) 276–284. doi:10.1080/08923973.2017.1344989.

[13]   J. Yan, Y. Chen, C. He, Z. Yang, C. Lü, X. Chen, Andrographolide induces cell cycle arrest and apoptosis in human rheumatoid arthritis fibroblast-like synoviocytes, Cell Biol. Toxicol. 28 (2012) 47–56. doi:10.1007/s10565-011-9204-8.

[14]   X. Ji, C. Li, Y. Ou, N. Li, K. Yuan, G. Yang, X. Chen, Z. Yang, B. Liu, W.W. Cheung, L. Wang, R. Huang, T. Lan, Andrographolide ameliorates diabetic nephropathy by attenuating hyperglycemia-mediated renal oxidative stress and inflammation via Akt/NF-κB pathway, Mol. Cell. Endocrinol. 437 (2016) 268–279. doi:10.1016/j.mce.2016.06.029.

[15]   S. Roy, R. Bhat, Effect of polyols on the structure and aggregation of recombinant human γ-Synuclein, an intrinsically disordered protein, Biochim. Biophys. Acta - Proteins Proteomics. 1866 (2018) 1029–1042. doi:10.1016/j.bbapap.2018.07.003.

[16]   S. Sen, S. Konar, B. Das, A. Pathak, S. Dhara, S. Dasgupta, S. DasGupta, Inhibition of fibrillation of human serum albumin through interaction with chitosan-based biocompatible silver nanoparticles, RSC Adv. 6 (2016) 43104–43115. doi:10.1039/C6RA05129D.

[17]   C. Basu, R. Sur, S-Allyl Cysteine Alleviates Hydrogen Peroxide Induced Oxidative Injury and Apoptosis through Upregulation of Akt/Nrf-2/HO-1 Signaling Pathway in

HepG2 Cells, Biomed Res. Int. 2018 (2018) 1–14. doi:10.1155/2018/3169431.

[18] T. Hossain, A. Saha, A. Mukherjee, Exploring molecular structural requirement for AChE inhibition through multi-chemometric and dynamics simulation analyses, J. Biomol. Struct. Dyn. 36 (2018) 1274–1285. doi:10.1080/07391102.2017.1320231.

[19] E. Harder, W. Damm, J. Maple, C. Wu, M. Reboul, J.Y. Xiang, L. Wang, D. Lupyan, M.K. Dahlgren, J.L. Knight, J.W. Kaus, D.S. Cerutti, G. Krilov, W.L. Jorgensen, R. Abel, R.A. Friesner, OPLS3: A Force Field Providing Broad Coverage of Drug-like Small Molecules and Proteins, J. Chem. Theory Comput. 12 (2016) 281–296. doi:10.1021/acs.jctc.5b00864.

[20] R.A. Friesner, R.B. Murphy, M.P. Repasky, L.L. Frye, J.R. Greenwood, T.A. Halgren, P.C. Sanschagrin, D.T. Mainz, Extra Precision Glide: Docking and Scoring Incorporating a Model of Hydrophobic Enclosure for Protein−Ligand Complexes, J. Med. Chem. 49 (2006) 6177–6196. doi:10.1021/jm0512560.

[21] L. Nicoud, M. Lattuada, A. Yates, M. Morbidelli, Impact of aggregate formation on the viscosity of protein solutions, Soft Matter. 11 (2015) 5513–5522. doi:10.1039/C5SM00513B.

[22] L. Yuan, M. Liu, B. Sun, J. Liu, X. Wei, Z. Wang, B. Wang, J. Han, Calorimetric and spectroscopic studies on the competitive behavior between (−)-epigallocatechin-3-gallate and 5-fluorouracil with human serum albumin, J. Mol. Liq. 248 (2017) 330–339. doi:10.1016/j.molliq.2017.10.049.

[23] N.A. Al-Shabib, J.M. Khan, M.A. Alsenaidy, A.M. Alsenaidy, M.S. Khan, F.M. Husain, M.R. Khan, M. Naseem, P. Sen, P. Alam, R.H. Khan, Unveiling the stimulatory effects of tartrazine on human and bovine serum albumin fibrillogenesis: Spectroscopic and microscopic study, Spectrochim. Acta - Part A Mol. Biomol. Spectrosc. 191 (2018) 116–124. doi:10.1016/j.saa.2017.09.062.

[24] T. Zheng, P. Cherubin, L. Cilenti, K. Teter, Q. Huo, A simple and fast method to study the hydrodynamic size difference of protein disulfide isomerase in oxidized and reduced form using gold nanoparticles and dynamic light scattering, Analyst. 141 (2016) 934–938. doi:10.1039/C5AN02248G.

[25] H. LeVine, Quantification of β-sheet amyloid fibril structures with thioflavin T, in:

Methods Enzymol., 1999: pp. 274–284. doi:10.1016/S0076-6879(99)09020-5.

[26]   R. Amini, R. Yazdanparast, S. Bahramikia, Apigenin reduces human insulin fibrillation in vitro and protects SK-N-MC cells against insulin amyloids, Int. J. Biol. Macromol. 60 (2013) 334–340. doi:10.1016/j.ijbiomac.2013.06.013.

[27]   S.K. Chaturvedi, M.K. Siddiqi, P. Alam, R.H. Khan, Protein misfolding and aggregation: Mechanism, factors and detection, Process Biochem. 51 (2016) 1183–1192. doi:10.1016/j.procbio.2016.05.015.

[28]   A. Varma, H. Padh, N. Shrivastava, Andrographolide: A New Plant-Derived Antineoplastic Entity on Horizon, Evidence-Based Complement. Altern. Med. 2011 (2011) 815390. doi:10.1093/ecam/nep135.

[29]   I. Khan, F. Khan, A. Farooqui, I.A. Ansari, Andrographolide Exhibits Anticancer Potential Against Human Colon Cancer Cells by Inducing Cell Cycle Arrest and Programmed Cell Death via Augmentation of Intracellular Reactive Oxygen Species Level, Nutr. Cancer. 70 (2018) 787–803. doi:10.1080/01635581.2018.1470649.

[30]   X. Wang, Y. Liu, L.L. He, B. Liu, S.Y. Zhang, X. Ye, J.J. Jing, J.F. Zhang, M. Gao, X. Wang, Spectroscopic investigation on the food components-drug interaction: The influence of flavonoids on the affinity of nifedipine to human serum albumin, Food Chem. Toxicol. 78 (2015) 42–51. doi:10.1016/j.fct.2015.01.026.

[31]   T.S. Banipal, N. Kaur, P.K. Banipal, Binding studies of caffeine and theophylline to bovine serum albumin: Calorimetric and spectroscopic approach, J. Mol. Liq. 223 (2016) 1048–1055. doi:10.1016/j.molliq.2016.09.034.

[32]   S. Roy, T.K. Das, Interaction of biosynthesized gold nanoparticles with BSA and CTDNA: A multi-spectroscopic approach, Polyhedron. 115 (2016) 111–118. doi:10.1016/j.poly.2016.05.002.

[33]   H.A. Benesi, J.H. Hildebrand, a Spectrophotometric Investigation of the Interaction of Iodine With Aromatic Hydrocarbons, J. Am. Chem. Soc. 71 (1949) 2703–2707. doi:10.1021/ja01176a030.

[34]   P. Singla, V. Luxami, K. Paul, Synthesis and in vitro evaluation of novel triazine analogues as anticancer agents and their interaction studies with bovine serum albumin, Eur. J. Med. Chem. 117 (2016) 59–69. doi:10.1016/j.ejmech.2016.03.088.

[35]   B. Yang, R. Liu, X. Hao, Y. Wu, J. Du, Effect of CdTe quantum dots size on the conformational changes of human serum albumin: Results of spectroscopy and isothermal titration calorimetry, Biol. Trace Elem. Res. 155 (2013) 150–158. doi:10.1007/s12011-013-9771-z.

[36]   S.N. Save, S. Choudhary, Effects of triphala and guggul aqueous extracts on inhibition of protein fibrillation and dissolution of preformed fibrils, RSC Adv. 7 (2017) 20460–20468. doi:10.1039/C6RA28440J.

[37]   M. Dockal, D.C. Carter, F. Rüker, Conformational transitions of the three recombinant domains of human serum albumin depending on pH, J. Biol. Chem. 275 (2000) 3042–3050. doi:10.1074/jbc.275.5.3042.

[38]   A. Micsonai, F. Wien, L. Kernya, Y.-H. Lee, Y. Goto, M. Réfrégiers, J. Kardos, Accurate secondary structure prediction and fold recognition for circular dichroism spectroscopy, Proc. Natl. Acad. Sci. 112 (2015) E3095–E3103. doi:10.1073/pnas.1500851112.

[39]   M. Wahlbom, X. Wang, V. Lindström, E. Carlemalm, M. Jaskolski, A. Grubb, Fibrillogenic oligomers of human cystatin C are formed by propagated domain swapping, J. Biol. Chem. 282 (2007) 18318–18326. doi:10.1074/jbc.M611368200.

[40]   B. Kost, M. Svyntkivska, M. Brzeziński, T. Makowski, E. Piorkowska, K. Rajkowska, A. Kunicka-Styczyńska, T. Biela, PLA/β-CD-based fibres loaded with quercetin as potential antibacterial dressing materials, Colloids Surfaces B Biointerfaces. 190 (2020) 110949. doi:10.1016/j.colsurfb.2020.110949.

[41]   F. Dalgaard, N.P. Bondonno, K. Murray, C.P. Bondonno, J.R. Lewis, K.D. Croft, C. Kyrø, G. Gislason, A. Scalbert, A. Cassidy, A. Tjønneland, K. Overvad, J.M. Hodgson, Associations between habitual flavonoid intake and hospital admissions for atherosclerotic cardiovascular disease: a prospective cohort study, Lancet Planet. Heal. 3 (2019) e450–e459. doi:10.1016/S2542-5196(19)30212-8.

[42]   M. Farazuddin, R. Mishra, Y. Jing, V. Srivastava, A.T. Comstock, U.S. Sajjan, Quercetin prevents rhinovirus-induced progression of lung disease in mice with COPD phenotype, PLoS One. 13 (2018) e0199612. doi:10.1371/journal.pone.0199612.

[43]   E. Gansukh, A. Nile, D.H. Kim, J.W. Oh, S.H. Nile, New insights into antiviral and

cytotoxic potential of quercetin and its derivatives – A biochemical perspective, Food Chem. 334 (2021) 127508. doi:10.1016/j.foodchem.2020.127508.

[44] C.S. Yang, J.M. Landau, M.-T. Huang, H.L. Newmark, INHIBITION OF CARCINOGENESIS BY DIETARY POLYPHENOLIC COMPOUNDS, Annu. Rev. Nutr. 21 (2001) 381–406. doi:10.1146/annurev.nutr.21.1.381.

[45] S.-M. Choi, B.C. Kim, Y.-H. Cho, K.-H. Choi, J. Chang, M.-S. Park, M.-K. Kim, K.-H. Cho, J.-K. Kim, Effects of Flavonoid Compounds on β-amyloid-peptide-induced Neuronal Death in Cultured Mouse Cortical Neurons, Chonnam Med. J. 50 (2014) 45–51. doi:10.4068/cmj.2014.50.2.45.

[46] A.M. Sabogal-Guáqueta, J.I. Muñoz-Manco, J.R. Ramírez-Pineda, M. Lamprea-Rodriguez, E. Osorio, G.P. Cardona-Gómez, The flavonoid quercetin ameliorates Alzheimer's disease pathology and protects cognitive and emotional function in aged triple transgenic Alzheimer's disease model mice, Neuropharmacology. 93 (2015) 134–145. doi:10.1016/j.neuropharm.2015.01.027.

[47] S.S. Karuppagounder, S.K. Madathil, M. Pandey, R. Haobam, U. Rajamma, K.P. Mohanakumar, Quercetin up-regulates mitochondrial complex-I activity to protect against programmed cell death in rotenone model of Parkinson's disease in rats, Neuroscience. 236 (2013) 136–148. doi:10.1016/j.neuroscience.2013.01.032.

[48] M.F. Mahmoud, N.A. Hassan, H.M. El Bassossy, A. Fahmy, Quercetin protects against diabetes-induced exaggerated vasoconstriction in rats: effect on low grade inflammation, PLoS One. 8 (2013) e63784–e63784. doi:10.1371/journal.pone.0063784.

[49] J. Dong, X. Zhang, L. Zhang, H.-X. Bian, N. Xu, B. Bao, J. Liu, Quercetin reduces obesity-associated ATM infiltration and inflammation in mice: a mechanism including AMPKα1/SIRT1, J. Lipid Res. 55 (2014) 363–374. doi:10.1194/jlr.M038786.

[50] K. Jiménez-Aliaga, P. Bermejo-Bescós, J. Benedí, S. Martín-Aragón, Quercetin and rutin exhibit antiamyloidogenic and fibril-disaggregating effects in vitro and potent antioxidant activity in APPswe cells, Life Sci. 89 (2011) 939–945. doi:10.1016/j.lfs.2011.09.023.

[51] H. Kim, B.-S. Park, K.-G. Lee, C.Y. Choi, S.S. Jang, Y.-H. Kim, S.-E. Lee, Effects of

Naturally Occurring Compounds on Fibril Formation and Oxidative Stress of β-Amyloid, J. Agric. Food Chem. 53 (2005) 8537–8541. doi:10.1021/jf051985c.

[52]   J.B. Wang, Y.M. Wang, C.M. Zeng, Quercetin inhibits amyloid fibrillation of bovine insulin and destabilizes preformed fibrils, Biochem. Biophys. Res. Commun. 415 (2011) 675–679. doi:10.1016/j.bbrc.2011.10.135.

[53]   R. Batra, H. Chan, G. Kamath, R. Ramprasad, M.J. Cherukara, S.K.R.S. Sankaranarayanan, Screening of Therapeutic Agents for COVID-19 Using Machine Learning and Ensemble Docking Studies, J. Phys. Chem. Lett. (2020) 7058–7065. doi:10.1021/acs.jpclett.0c02278.

[54]   M.K. Siddiqi, P. Alam, S.K. Chaturvedi, S. Nusrat, Y.E. Shahein, R.H. Khan, Attenuation of amyloid fibrillation in presence of Warfarin: A biophysical investigation, Int. J. Biol. Macromol. 95 (2017) 713–718. doi:10.1016/j.ijbiomac.2016.11.110.

[55]   P. Alam, M.K. Siddiqi, S. Malik, S.K. Chaturvedi, M. Uddin, R.H. Khan, Elucidating the inhibitory potential of Vitamin A against fibrillation and amyloid associated cytotoxicity, Int. J. Biol. Macromol. 129 (2019) 333–338. doi:10.1016/j.ijbiomac.2019.01.134.

[56]   M. Groenning, L. Olsen, M. van de Weert, J.M. Flink, S. Frokjaer, F.S. Jørgensen, Study on the binding of Thioflavin T to β-sheet-rich and non-β-sheet cavities, J. Struct. Biol. 158 (2007) 358–369. doi:10.1016/j.jsb.2006.12.010.

[57]   M.R.H. Krebs, E.H.C. Bromley, A.M. Donald, The binding of thioflavin-T to amyloid fibrils: localisation and implications, J. Struct. Biol. 149 (2005) 30–37. doi:10.1016/j.jsb.2004.08.002.

[58]   N.K. Ramesh, S. Sudhakar, E. Mani, Modeling of the Inhibitory Effect of Nanoparticles on Amyloid β Fibrillation, Langmuir. 34 (2018) 4004–4012. doi:10.1021/acs.langmuir.8b00388.

[59]   M.K. Siddiqi, N. Majid, S. Malik, P. Alam, R.H. Khan, Amyloid Oligomers, Protofibrils and Fibrils, in: J.R. Harris, J. Marles-Wright (Eds.), Springer International Publishing, Cham, 2019: pp. 471–503. doi:10.1007/978-3-030-28151-9_16.

[60]   W.F. Bhat, I.A. Bhat, S.A. Bhat, B. Bano, In vitro disintegration of goat brain cystatin

fibrils using conventional and gemini surfactants: Putative therapeutic intervention in amyloidoses, Int. J. Biol. Macromol. 93 (2016) 493–500. doi:10.1016/j.ijbiomac.2016.08.082.

[61] R. Hornedo-Ortega, M.A. Álvarez-Fernández, A.B. Cerezo, T. Richard, A.M. Troncoso, M.C. Garcia-Parrilla, Protocatechuic Acid: Inhibition of Fibril Formation, Destabilization of Preformed Fibrils of Amyloid-β and α-Synuclein, and Neuroprotection, J. Agric. Food Chem. 64 (2016) 7722–7732. doi:10.1021/acs.jafc.6b03217.

[62] Z. Yang, C. Ge, J. Liu, Y. Chong, Z. Gu, C.A. Jimenez-Cruz, Z. Chai, R. Zhou, Destruction of amyloid fibrils by graphene through penetration and extraction of peptides, Nanoscale. 7 (2015) 18725–18737. doi:10.1039/C5NR01172H.

[63] J. Juárez, M. Alatorre-Meda, A. Cambón, A. Topete, S. Barbosa, P. Taboada, V. Mosquera, Hydration effects on the fibrillation process of a globular protein: the case of human serum albumin, Soft Matter. 8 (2012) 3608–3619. doi:10.1039/C2SM06762E.

[64] R. Singla, S.M.S. Abidi, A.I. Dar, A. Acharya, Inhibition of Glycation-Induced Aggregation of Human Serum Albumin by Organic–Inorganic Hybrid Nanocomposites of Iron Oxide-Functionalized Nanocellulose, ACS Omega. 4 (2019) 14805–14819. doi:10.1021/acsomega.9b01392.

[65] S. Roy, An insight of binding interaction between Tryptophan, Tyrosine and Phenylalanine separately with green gold nanoparticles by fluorescence quenching method, Optik (Stuttg). 138 (2017) 280–288. doi:10.1016/j.ijleo.2017.03.057.

[66] Z. Omidvar, K. Parivar, H. Sanee, Z. Amiri-Tehranizadeh, A. Baratian, M.R. Saberi, A. Asoodeh, J. Chamani, Investigations with Spectroscopy, Zeta Potential and Molecular Modeling of the Non-Cooperative Behaviour Between Cyclophosphamide Hydrochloride and Aspirin upon Interaction with Human Serum Albumin: Binary and Ternary Systems from the View Point of Multi-, J. Biomol. Struct. Dyn. 29 (2011) 181–206. doi:10.1080/07391102.2011.10507382.

[67] Q. Wang, C. Huang, M. Jiang, Y. Zhu, J. Wang, J. Chen, J. Shi, Binding interaction of atorvastatin with bovine serum albumin: Spectroscopic methods and molecular docking, Spectrochim. Acta Part A Mol. Biomol. Spectrosc. 156 (2016) 155–163.

doi:10.1016/j.saa.2015.12.003.

[68] J.C.N. Santos, I.M. da Silva, T.C. Braga, Â. de Fátima, I.M. Figueiredo, J.C.C. Santos, Thimerosal changes protein conformation and increase the rate of fibrillation in physiological conditions: Spectroscopic studies using bovine serum albumin (BSA), Int. J. Biol. Macromol. 113 (2018) 1032–1040. doi:https://doi.org/10.1016/j.ijbiomac.2018.02.116.

[69] S. Zargar, S. Alamery, A.H. Bakheit, T.A. Wani, Poziotinib and bovine serum albumin binding characterization and influence of quercetin, rutin, naringenin and sinapic acid on their binding interaction, Spectrochim. Acta Part A Mol. Biomol. Spectrosc. 235 (2020) 118335. doi:10.1016/j.saa.2020.118335.

[70] M. Faramarzian, S. Bahramikia, M. Dehghan Shasaltaneh, In vitro investigation of the effect of mesalazine on amyloid fibril formation of hen egg-white lysozyme and defibrillation lysozyme fibrils, Eur. J. Pharmacol. 874 (2020) 173011. doi:10.1016/j.ejphar.2020.173011.

[71] F. Mohammadi, M. Moeeni, A. Mahmudian, L. Hassani, Inhibition of amyloid fibrillation of lysozyme by bisdemethoxycurcumin and diacetylbisdemethoxycurcumin, Biophys. Chem. 235 (2018) 56–65. doi:10.1016/j.bpc.2018.02.005.

[72] J.T. Meijer, M. Roeters, V. Viola, D.W.P.M. Löwik, G. Vriend, J.C.M. van Hest, Stabilization of Peptide Fibrils by Hydrophobic Interaction, Langmuir. 23 (2007) 2058–2063. doi:10.1021/la0625345.

[73] A.W. Boots, J.M. Balk, A. Bast, G.R.M.M. Haenen, The reversibility of the glutathionyl-quercetin adduct spreads oxidized quercetin-induced toxicity, Biochem. Biophys. Res. Commun. 338 (2005) 923–929. doi:10.1016/j.bbrc.2005.10.031.

[74] A.B. Siegel, J. Stebbing, Milk thistle: early seeds of potential, Lancet Oncol. 14 (2013) 929–930. doi:10.1016/S1470-2045(13)70414-5.

[75] R. Saller, R. Meier, R. Brignoli, The Use of Silymarin in the Treatment of Liver Diseases, Drugs. 61 (2001) 2035–2063. doi:10.2165/00003495-200161140-00003.

[76] D. Delmas, Silymarin and Derivatives: From Biosynthesis to Health Benefits, Mol. . 25 (2020). doi:10.3390/molecules25102415.

[77]   J.I. Lee, M. Narayan, J.S. Barrett, Analysis and comparison of active constituents in commercial standardized silymarin extracts by liquid chromatography–electrospray ionization mass spectrometry, J. Chromatogr. B. 845 (2007) 95–103. doi:10.1016/j.jchromb.2006.07.063.

[78]   A. Federico, M. Dallio, M. Masarone, A.G. Gravina, R. Di Sarno, C. Tuccillo, V. Cossiga, S. Lama, P. Stiuso, F. Morisco, M. Persico, C. Loguercio, Evaluation of the Effect Derived from Silybin with Vitamin D and Vitamin E Administration on Clinical, Metabolic, Endothelial Dysfunction, Oxidative Stress Parameters, and Serological Worsening Markers in Nonalcoholic Fatty Liver Disease Patients, Oxid. Med. Cell. Longev. 2019 (2019) 8742075. doi:10.1155/2019/8742075.

[79]   L. Tao, X. Qu, Y. Zhang, Y. Song, S. Zhang, Prophylactic Therapy of Silymarin (Milk Thistle) on Antituberculosis Drug-Induced Liver Injury: A Meta-Analysis of Randomized Controlled Trials, Can. J. Gastroenterol. Hepatol. 2019 (2019) 3192351. doi:10.1155/2019/3192351.

[80]   P. Trouillas, P. Marsal, A. Svobodová, J. Vostálová, R. Gažák, J. Hrbáč, P. Sedmera, V. Křen, R. Lazzaroni, J.-L. Duroux, D. Walterová, Mechanism of the Antioxidant Action of Silybin and 2,3-Dehydrosilybin Flavonolignans: A Joint Experimental and Theoretical Study, J. Phys. Chem. A. 112 (2008) 1054–1063. doi:10.1021/jp075814h.

[81]   M. Trappoliere, A. Caligiuri, M. Schmid, C. Bertolani, P. Failli, F. Vizzutti, E. Novo, C. di Manzano, F. Marra, C. Loguercio, M. Pinzani, Silybin, a component of sylimarin, exerts anti-inflammatory and anti-fibrogenic effects on human hepatic stellate cells, J. Hepatol. 50 (2009) 1102–1111. doi:10.1016/j.jhep.2009.02.023.

[82]   S. Das, P. Roy, R. Pal, R.G. Auddy, A.S. Chakraborti, A. Mukherjee, Engineered silybin nanoparticles educe efficient control in experimental diabetes, PLoS One. 9 (2014) e101818–e101818. doi:10.1371/journal.pone.0101818.

[83]   P. Liu, L. Cui, B. Liu, W. Liu, T. Hayashi, K. Mizuno, S. Hattori, Y. Ushiki-Kaku, S. Onodera, T. Ikejima, Silibinin ameliorates STZ-induced impairment of memory and learning by up- regulating insulin signaling pathway and attenuating apoptosis, Physiol. Behav. 213 (2020) 112689. doi:10.1016/j.physbeh.2019.112689.

[84]   W. Sherman, H.S. Beard, R. Farid, Use of an Induced Fit Receptor Structure in Virtual Screening, Chem. Biol. Drug Des. 67 (2006) 83–84. doi:10.1111/j.1747-

0285.2005.00327.x.

[85]   J. Juárez, P. Taboada, V. Mosquera, Existence of different structural intermediates on the fibrillation pathway of human serum albumin, Biophys. J. 96 (2009) 2353–2370. doi:10.1016/j.bpj.2008.12.3901.

[86]   Z. Dvořák, P. Kosina, D. Walterová, V. Šimánek, P. Bachleda, J. Ulrichová, Primary cultures of human hepatocytes as a tool in cytotoxicity studies: cell protection against model toxins by flavonolignans obtained from Silybum marianum, Toxicol. Lett. 137 (2003) 201–212. doi:10.1016/S0378-4274(02)00406-X.

[87]   M. Gharagozloo, Z. Khoshdel, Z. Amirghofran, The effect of an iron (III) chelator, silybin, on the proliferation and cell cycle of Jurkat cells: A comparison with desferrioxamine, Eur. J. Pharmacol. 589 (2008) 1–7. doi:10.1016/j.ejphar.2008.03.059.

[88]   L. Radko, W. Cybulski, W. Rzeski, The Protective Effect of Silybin against Lasalocid Cytotoxic Exposure on Chicken and Rat Cell Lines, Biomed Res. Int. 2013 (2013) 1–8. doi:10.1155/2013/783519.

[89]   S.P. Bhimaneni, V. Bhati, S. Bhosale, A. Kumar, Investigates interaction between abscisic acid and bovine serum albumin using various spectroscopic and in-silico techniques, J. Mol. Struct. 1224 (2021) 129018. doi:10.1016/j.molstruc.2020.129018.

[90]   F. Chiti, C.M. Dobson, Protein Misfolding, Functional Amyloid, and Human Disease, Annu. Rev. Biochem. 75 (2006) 333–366. doi:10.1146/annurev.biochem.75.101304.123901.

[91]   K. Niikura, N. Iyo, Y. Matsuo, H. Mitomo, K. Ijiro, Sub-100 nm Gold Nanoparticle Vesicles as a Drug Delivery Carrier enabling Rapid Drug Release upon Light Irradiation, ACS Appl. Mater. Interfaces. 5 (2013) 3900–3907. doi:10.1021/am400590m.

[92]   K.B. Johnsen, M. Bak, F. Melander, M.S. Thomsen, A. Burkhart, P.J. Kempen, T.L. Andresen, T. Moos, Modulating the antibody density changes the uptake and transport at the blood-brain barrier of both transferrin receptor-targeted gold nanoparticles and liposomal cargo, J. Control. Release. 295 (2019) 237–249. doi:10.1016/j.jconrel.2019.01.005.

[93]   A.L. Lira, R.S. Ferreira, R.J.S. Torquato, H. Zhao, M.L. V Oliva, S.A. Hassan, P.

Schuck, A.A. Sousa, Binding kinetics of ultrasmall gold nanoparticles with proteins, Nanoscale. 10 (2018) 3235–3244. doi:10.1039/C7NR06810G.

[94]   Y. Zhao, W. Zhao, Y.C. Lim, T. Liu, Salinomycin-Loaded Gold Nanoparticles for Treating Cancer Stem Cells by Ferroptosis-Induced Cell Death, Mol. Pharm. 16 (2019) 2532–2539. doi:10.1021/acs.molpharmaceut.9b00132.

[95]   T. Cui, J.-J. Liang, H. Chen, D.-D. Geng, L. Jiao, J.-Y. Yang, H. Qian, C. Zhang, Y. Ding, Performance of Doxorubicin-Conjugated Gold Nanoparticles: Regulation of Drug Location, ACS Appl. Mater. Interfaces. 9 (2017) 8569–8580. doi:10.1021/acsami.6b16669.

[96]   P. Narayanasamy, B.L. Switzer, B.E. Britigan, Prolonged-acting, Multi-targeting Gallium Nanoparticles Potently Inhibit Growth of Both HIV and Mycobacteria in Co-Infected Human Macrophages, Sci. Rep. 5 (2015) 8824. doi:10.1038/srep08824.

[97]   J. das Neves, M.M. Amiji, M.F. Bahia, B. Sarmento, Nanotechnology-based systems for the treatment and prevention of HIV/AIDS, Adv. Drug Deliv. Rev. 62 (2010) 458–477. doi:10.1016/j.addr.2009.11.017.

[98]   B.J. Edagwa, D. Guo, P. Puligujja, H. Chen, J. McMillan, X. Liu, H.E. Gendelman, P. Narayanasamy, Long-acting antituberculous therapeutic nanoparticles target macrophage endosomes, FASEB J. 28 (2014) 5071–5082. doi:10.1096/fj.14-255786.

[99]   H. Skaat, R. Chen, I. Grinberg, S. Margel, Engineered Polymer Nanoparticles Containing Hydrophobic Dipeptide for Inhibition of Amyloid-β Fibrillation, Biomacromolecules. 13 (2012) 2662–2670. doi:10.1021/bm3011177.

[100]   C. Cabaleiro-Lago, O. Szczepankiewicz, S. Linse, The effect of nanoparticles on amyloid aggregation depends on the protein stability and intrinsic aggregation rate, Langmuir. 28 (2012) 1852–1857. doi:10.1021/la203078w.

[101]   N.G. Khlebtsov, Determination of size and concentration of gold nanoparticles from extinction spectra, Anal. Chem. 80 (2008) 6620–6625. doi:10.1021/ac800834n.

[102]   V. Sanna, N. Pala, G. Dessì, P. Manconi, A. Mariani, S. Dedola, M. Rassu, C. Crosio, C. Iaccarino, M. Sechi, Single-step green synthesis and characterization of gold-conjugated polyphenol nanoparticles with antioxidant and biological activities, Int. J. Nanomedicine. 9 (2014) 4935–4951. doi:10.2147/IJN.S70648.

[103] N. Nenadis, O. Lazaridou, M.Z. Tsimidou, Use of Reference Compounds in Antioxidant Activity Assessment, J. Agric. Food Chem. 55 (2007) 5452–5460. doi:10.1021/jf070473q.

[104] S. Goy-López, J. Juárez, M. Alatorre-Meda, E. Casals, V.F. Puntes, P. Taboada, V. Mosquera, Physicochemical Characteristics of Protein–NP Bioconjugates: The Role of Particle Curvature and Solution Conditions on Human Serum Albumin Conformation and Fibrillogenesis Inhibition, Langmuir. 28 (2012) 9113–9126. doi:10.1021/la300402w.

[105] S. Sudhakar, P. Kalipillai, P.B. Santhosh, E. Mani, Role of Surface Charge of Inhibitors on Amyloid Beta Fibrillation, J. Phys. Chem. C. 121 (2017) 6339–6348. doi:10.1021/acs.jpcc.6b12307.

[106] J. Polte, R. Erler, A.F. Thünemann, S. Sokolov, T.T. Ahner, K. Rademann, F. Emmerling, R. Kraehnert, Nucleation and Growth of Gold Nanoparticles Studied via in situ Small Angle X-ray Scattering at Millisecond Time Resolution, ACS Nano. 4 (2010) 1076–1082. doi:10.1021/nn901499c.

[107] S.J. Hwang, S.H. Jun, Y. Park, S.-H. Cha, M. Yoon, S. Cho, H.-J. Lee, Y. Park, Green synthesis of gold nanoparticles using chlorogenic acid and their enhanced performance for inflammation, Nanomedicine Nanotechnology, Biol. Med. 11 (2015) 1677–1688. doi:10.1016/j.nano.2015.05.002.

[108] J. Turkevich, P.C. Stevenson, J. Hillier, A study of the nucleation and growth processes in the synthesis of colloidal gold, Discuss. Faraday Soc. 11 (1951) 55–75. doi:10.1039/DF9511100055.

[109] J. Kimling, M. Maier, B. Okenve, V. Kotaidis, H. Ballot, A. Plech, Turkevich Method for Gold Nanoparticle Synthesis Revisited, J. Phys. Chem. B. 110 (2006) 15700–15707. doi:10.1021/jp061667w.

[110] J. Piella, N.G. Bastús, V. Puntes, Size-Controlled Synthesis of Sub-10-nanometer Citrate-Stabilized Gold Nanoparticles and Related Optical Properties., Chem. Mater. 28 (2016) 1066–1075. doi:10.1021/acs.chemmater.5b04406.

[111] S. Das, P. Roy, S. Mondal, T. Bera, A. Mukherjee, One pot synthesis of gold nanoparticles and application in chemotherapy of wild and resistant type visceral

leishmaniasis, Colloids Surfaces B Biointerfaces. 107 (2013) 27–34. doi:10.1016/j.colsurfb.2013.01.061.

[112] A. Halder, S. Das, T. Bera, A. Mukherjee, Rapid synthesis for monodispersed gold nanoparticles in kaempferol and anti-leishmanial efficacy against wild and drug resistant strains, RSC Adv. 7 (2017) 14159–14167. doi:10.1039/c6ra28632a.

[113] T.C. Prathna, N. Chandrasekaran, A.M. Raichur, A. Mukherjee, Biomimetic synthesis of silver nanoparticles by Citrus limon (lemon) aqueous extract and theoretical prediction of particle size, Colloids Surfaces B Biointerfaces. 82 (2011) 152–159. doi:10.1016/j.colsurfb.2010.08.036.

[114] S.H. De Paoli Lacerda, J.J. Park, C. Meuse, D. Pristinski, M.L. Becker, A. Karim, J.F. Douglas, Interaction of gold nanoparticles with common human blood proteins, ACS Nano. 4 (2010) 365–379. doi:10.1021/nn9011187.

[115] S. Das, P. Roy, R.G. Auddy, A. Mukherjee, Silymarin nanoparticle prevents paracetamol-induced hepatotoxicity., Int. J. Nanomedicine. 6 (2011) 1291–1301. doi:10.2147/IJN.S15160.

[116] T. Kongpichitchoke, J.-L. Hsu, T.-C. Huang, Number of Hydroxyl Groups on the B-Ring of Flavonoids Affects Their Antioxidant Activity and Interaction with Phorbol Ester Binding Site of PKCδ C1B Domain: In Vitro and in Silico Studies, J. Agric. Food Chem. 63 (2015) 4580–4586. doi:10.1021/acs.jafc.5b00312.

[117] K. Pyrzynska, A. Pękal, Application of free radical diphenylpicrylhydrazyl (DPPH) to estimate the antioxidant capacity of food samples, Anal. Methods. 5 (2013) 4288–4295. doi:10.1039/c3ay40367j.

[118] S. Sen, S. Dasgupta, S. DasGupta, Does Surface Chirality of Gold Nanoparticles Affect Fibrillation of HSA?, J. Phys. Chem. C. 121 (2017) 18935–18946. doi:10.1021/acs.jpcc.7b05354.

[119] M.K. Siddiqi, P. Alam, S.K. Chaturvedi, R.H. Khan, Anti-amyloidogenic behavior and interaction of Diallylsulfide with Human Serum Albumin, Int. J. Biol. Macromol. 92 (2016) 1220–1228. doi:10.1016/j.ijbiomac.2016.08.035.

[120] S. Bag, R. Mitra, S. DasGupta, S. Dasgupta, Inhibition of Human Serum Albumin Fibrillation by Two-Dimensional Nanoparticles, J. Phys. Chem. B. 121 (2017) 5474–

5482. doi:10.1021/acs.jpcb.7b01289.

[121] S. Sen, S. Konar, A. Pathak, S. Dasgupta, S. Dasgupta, Effect of functionalized magnetic MnFe2O4 nanoparticles on fibrillation of human serum albumin, J. Phys. Chem. B. 118 (2014) 11667–11676. doi:10.1021/jp507902y.

[122] N. Xiong, X.Y. Dong, J. Zheng, F.F. Liu, Y. Sun, Design of LVFFARK and LVFFARK-Functionalized Nanoparticles for Inhibiting Amyloid β-Protein Fibrillation and Cytotoxicity, ACS Appl. Mater. Interfaces. 7 (2015) 5650–5662. doi:10.1021/acsami.5b00915.

[123] M. Jafari, F. Rokhbakhsh-Zamin, M. Shakibaie, M.H. Moshafi, A. Ameri, H.R. Rahimi, H. Forootanfar, Cytotoxic and antibacterial activities of biologically synthesized gold nanoparticles assisted by Micrococcus yunnanensis strain J2, Biocatal. Agric. Biotechnol. 15 (2018) 245–253. doi:10.1016/j.bcab.2018.06.014.

[124] N.M. Schaeublin, L.K. Braydich-Stolle, A.M. Schrand, J.M. Miller, J. Hutchison, J.J. Schlager, S.M. Hussain, Surface charge of gold nanoparticles mediates mechanism of toxicity, Nanoscale. 3 (2011) 410–420. doi:10.1039/c0nr00478b.

[125] Y. Luo, Q. Wang, Y. Zhang, Biopolymer-Based Nanotechnology Approaches To Deliver Bioactive Compounds for Food Applications: A Perspective on the Past, Present, and Future, J. Agric. Food Chem. (2020) acs.jafc.0c00277. doi:10.1021/acs.jafc.0c00277.

[126] J. Karlsson, H.J. Vaughan, J.J. Green, Biodegradable Polymeric Nanoparticles for Therapeutic Cancer Treatments, Annu. Rev. Chem. Biomol. Eng. 9 (2018) 105–127. doi:10.1146/annurev-chembioeng-060817-084055.

[127] S. Alqahtani, L. Simon, C.E. Astete, A. Alayoubi, P.W. Sylvester, S. Nazzal, Y. Shen, Z. Xu, A. Kaddoumi, C.M. Sabliov, Cellular uptake, antioxidant and antiproliferative activity of entrapped α-tocopherol and γ-tocotrienol in poly (lactic-co-glycolic) acid (PLGA) and chitosan covered PLGA nanoparticles (PLGA-Chi), J. Colloid Interface Sci. 445 (2015) 243–251. doi:10.1016/j.jcis.2014.12.083.

[128] X. Xu, W. Ho, X. Zhang, N. Bertrand, O. Farokhzad, Cancer nanomedicine: From targeted delivery to combination therapy, Trends Mol. Med. 21 (2015) 223–232. doi:10.1016/j.molmed.2015.01.001.

[129] L.M. Ickenstein, P. Garidel, Lipid-based nanoparticle formulations for small molecules and RNA drugs, Expert Opin. Drug Deliv. 16 (2019) 1205–1226. doi:10.1080/17425247.2019.1669558.

[130] N.M. Gulati, P.L. Stewart, N.F. Steinmetz, Bioinspired Shielding Strategies for Nanoparticle Drug Delivery Applications, Mol. Pharm. 15 (2018) 2900–2909. doi:10.1021/acs.molpharmaceut.8b00292.

[131] S. Wilhelm, A.J. Tavares, Q. Dai, S. Ohta, J. Audet, H.F. Dvorak, W.C.W. Chan, Analysis of nanoparticle delivery to tumours, Nat. Rev. Mater. 1 (2016) 16014. doi:10.1038/natrevmats.2016.14.

[132] V. Mirshafiee, M. Mahmoudi, K. Lou, J. Cheng, M.L. Kraft, Protein corona significantly reduces active targeting yield, Chem. Commun. 49 (2013) 2557–2559. doi:10.1039/c3cc37307j.

[133] T. Cedervall, I. Lynch, S. Lindman, T. Berggard, E. Thulin, H. Nilsson, K.A. Dawson, S. Linse, Understanding the nanoparticle-protein corona using methods to quantify exchange rates and affinities of proteins for nanoparticles, Proc. Natl. Acad. Sci. 104 (2007) 2050–2055. doi:10.1073/pnas.0608582104.

[134] F. Bonaccorso, M. Zerbetto, A.C. Ferrari, V. Amendola, Sorting nanoparticles by centrifugal fields in clean media, J. Phys. Chem. C. 117 (2013) 13217–13229. doi:10.1021/jp400599g.

[135] G.M. Xiong, S. Yuan, J.K. Wang, A.T. Do, N.S. Tan, K.S. Yeo, C. Choong, Imparting electroactivity to polycaprolactone fibers with heparin-doped polypyrrole: Modulation of hemocompatibility and inflammatory responses, Acta Biomater. 23 (2015) 240–249. doi:10.1016/j.actbio.2015.05.003.

[136] M.J. Hajipour, S. Laurent, A. Aghaie, F. Rezaee, M. Mahmoudi, Personalized protein coronas: a "key" factor at the nanobiointerface, Biomater. Sci. 2 (2014) 1210. doi:10.1039/C4BM00131A.

[137] M.M. Bradford, A rapid and sensitive method for the quantitation of microgram quantities of protein utilizing the principle of protein-dye binding, Anal. Biochem. 72 (1976) 248–254. doi:10.1016/0003-2697(76)90527-3.

[138] L. Fu, Y. Sun, L. Ding, Y. Wang, Z. Gao, Z. Wu, S. Wang, W. Li, Y. Bi, Mechanism

evaluation of the interactions between flavonoids and bovine serum albumin based on multi-spectroscopy, molecular docking and Q-TOF HR-MS analyses, Food Chem. 203 (2016) 150–157. doi:10.1016/j.foodchem.2016.01.105.

[139] S.K. Vaishanav, K. Chandraker, J. Korram, R. Nagwanshi, K.K. Ghosh, M.L. Satnami, Protein nanoparticle interaction: A spectrophotometric approach for adsorption kinetics and binding studies, J. Mol. Struct. 1117 (2016) 300–310. doi:10.1016/j.molstruc.2016.03.087.

[140] S. Chakraborti, S. Sarwar, P. Chakrabarti, The effect of the binding of zno nanoparticle on the structure and stability of α-lactalbumin: A comparative study, J. Phys. Chem. B. 117 (2013) 13397–13408. doi:10.1021/jp404411b.

[141] G.J. Pillai, M.M. Greeshma, D. Menon, Impact of poly(lactic-co-glycolic acid) nanoparticle surface charge on protein, cellular and haematological interactions, Colloids Surfaces B Biointerfaces. 136 (2015) 1058–1066. doi:10.1016/j.colsurfb.2015.10.047.

[142] J. Juárez, S.G. López, A. Cambón, P. Taboada, V. Mosquera, Influence of electrostatic interactions on the fibrillation process of human serum albumin., J. Phys. Chem. B. 113 (2009) 10521–10529. doi:10.1021/jp902224d.

[143] Y. Il Chung, J.C. Kim, Y.H. Kim, G. Tae, S.Y. Lee, K. Kim, I.C. Kwon, The effect of surface functionalization of PLGA nanoparticles by heparin- or chitosan-conjugated Pluronic on tumor targeting, J. Control. Release. 143 (2010) 374–382. doi:10.1016/j.jconrel.2010.01.017.

[144] A. Albanese, P.S. Tang, W.C.W. Chan, The Effect of Nanoparticle Size, Shape, and Surface Chemistry on Biological Systems, Annu. Rev. Biomed. Eng. 14 (2012) 1–16. doi:10.1146/annurev-bioeng-071811-150124.

[145] D.F. Moyano, K. Saha, G. Prakash, B. Yan, H. Kong, M. Yazdani, V.M. Rotello, Fabrication of Corona-Free Nanoparticles with Tunable Hydrophobicity, ACS Nano. 8 (2014) 6748–6755. doi:10.1021/nn5006478.

[146] C. Gunawan, M. Lim, C.P. Marquis, R. Amal, Nanoparticle-protein corona complexes govern the biological fates and functions of nanoparticles, J. Mater. Chem. B. 2 (2014) 2060–2083. doi:10.1039/c3tb21526a.

[147] M.M. Yallapu, N. Chauhan, S.F. Othman, V. Khalilzad-sharghi, M.C. Ebeling, S. Khan, M. Jaggi, S.C. Chauhan, Implications of protein corona on physico-chemical and biological properties of magnetic nanoparticles, Biomaterials. 46 (2015) 1–12. doi:10.1016/j.biomaterials.2014.12.045.

[148] D.H. Napper, A. Netschey, Studies of the steric stabilization of colloidal particles, J. Colloid Interface Sci. 37 (1971) 528–535. doi:10.1016/0021-9797(71)90330-4.

[149] J.F. Carpenter, J.H. Crowe, An infrared spectroscopic study of the interactions of carbohydrates with dried proteins, Biochemistry. 28 (1989) 3916–3922. doi:10.1021/bi00435a044.

[150] D.M. Mann, E. Romm, M. Migliorini, Delineation of the glycosaminoglycan-binding site in the human inflammatory response protein lactoferrin, J. Biol. Chem. 269 (1994) 23661–23667.

[151] M. Lundqvist, J. Stigler, T. Cedervall, T. Berggård, M.B. Flanagan, I. Lynch, G. Elia, K. Dawson, The evolution of the protein corona around nanoparticles: A test study, ACS Nano. 5 (2011) 7503–7509. doi:10.1021/nn202458g.

[152] S. Milani, F. Baldelli Bombelli, A.S. Pitek, K.A. Dawson, J. Rädler, Reversible versus irreversible binding of transferrin to polystyrene nanoparticles: Soft and hard corona, ACS Nano. 6 (2012) 2532–2541. doi:10.1021/nn204951s.

[153] S. Winzen, S. Schoettler, G. Baier, C. Rosenauer, V. Mailaender, K. Landfester, K. Mohr, Complementary analysis of the hard and soft protein corona: sample preparation critically effects corona composition, Nanoscale. 7 (2015) 2992–3001. doi:10.1039/C4NR05982D.

[154] C.D. Hoemann, D. Fong, Immunological responses to chitosan for biomedical applications, Elsevier, 2016. doi:10.1016/B978-0-08-100230-8.00003-0.

[155] N. Weber, H.P. Wendel, G. Ziemer, Hemocompatibility of heparin-coated surfaces and the role of selective plasma protein adsorption, Biomaterials. 23 (2002) 429–439. doi:10.1016/S0142-9612(01)00122-3.

[156] P.D. Gorevic, T.T. Casey, W.J. Stone, C.R. DiRaimondo, F.C. Prelli, B. Frangione, Beta-2 microglobulin is an amyloidogenic protein in man, J. Clin. Invest. 76 (1985) 2425–2429. doi:10.1172/JCI112257.

[157] J. Lozier, N. Takahashi, F.W. Putnam, Complete amino acid sequence of human plasma beta 2-glycoprotein I., Proc. Natl. Acad. Sci. U. S. A. 81 (1984) 3640–4. doi:10.1073/pnas.81.12.3640.

[158] P. Taboada, S. Barbosa, E. Castro, V. Mosquera, Amyloid fibril formation and other aggregate species formed by human serum albumin association., J. Phys. Chem. B. 110 (2006) 20733–20736. doi:10.1021/jp064861r.

[159] J.S. Gebauer, M. Malissek, S. Simon, S.K. Knauer, M. Maskos, R.H. Stauber, W. Peukert, L. Treuel, Impact of the Nanoparticle–Protein Corona on Colloidal Stability and Protein Structure, Langmuir. 28 (2012) 9673–9679. doi:10.1021/la301104a.

[160] L. Bekale, D. Agudelo, H. a. Tajmir-Riahi, Effect of polymer molecular weight on chitosan-protein interaction, Colloids Surfaces B Biointerfaces. 125 (2015) 309–317. doi:10.1016/j.colsurfb.2014.11.037.

[161] E. Bazar, R. Jelinek, Divergent Heparin-Induced Fibrillation Pathways of a Prion Amyloidogenic Determinant, ChemBioChem. 11 (2010) 1997–2002. doi:10.1002/cbic.201000207.

[162] J.P. Solomon, S. Bourgault, E.T. Powers, J.W. Kelly, Heparin Binds 8 kDa Gelsolin Cross-β-Sheet Oligomers and Accelerates Amyloidogenesis by Hastening Fibril Extension, Biochemistry. 50 (2011) 2486–2498. doi:10.1021/bi101905n.

[163] T. Shacham, N. Sharma, G.Z. Lederkremer, Protein Misfolding and ER Stress in Huntington's Disease, Front. Mol. Biosci. 6 (2019) 20. doi:10.3389/fmolb.2019.00020.

[164] I.O. Kazaz, S. Demir, E. Yulug, F. Colak, A. Bodur, S.O. Yaman, E. Karaguzel, A. Mentese, N-acetylcysteine protects testicular tissue against ischemia/reperfusion injury via inhibiting endoplasmic reticulum stress and apoptosis, J. Pediatr. Urol. 15 (2019) 253.e1-253.e8. doi:10.1016/j.jpurol.2019.02.005.

[165] K. Kimura, H. Jin, M. Ogawa, T. Aoe, Dysfunction of the ER chaperone BiP accelerates the renal tubular injury, Biochem. Biophys. Res. Commun. 366 (2008) 1048–1053. doi:10.1016/j.bbrc.2007.12.098.

[166] W. Huang, J. Zhang, M. Washington, J. Liu, J.M. Parant, G. Lozano, D.D. Moore, Xenobiotic Stress Induces Hepatomegaly and Liver Tumors via the Nuclear Receptor Constitutive Androstane Receptor, Mol. Endocrinol. 19 (2005) 1646–1653.

doi:10.1210/me.2004-0520.

[167] H. Malhi, R.J. Kaufman, Endoplasmic reticulum stress in liver disease, J. Hepatol. 54 (2011) 795–809. doi:10.1016/j.jhep.2010.11.005.

[168] A. Rab, R. Bartoszewski, A. Jurkuvenaite, J. Wakefield, J.F. Collawn, Z. Bebők, Endoplasmic reticulum stress and the unfolded protein response regulate genomic cystic fibrosis transmembrane conductance regulator expression, Am. J. Physiol. Physiol. 292 (2007) C756–C766. doi:10.1152/ajpcell.00391.2006.

[169] C. Ji, N. Kaplowitz, Betaine decreases hyperhomocysteinemia, endoplasmic reticulum stress, and liver injury in alcohol-fed mice, Gastroenterology. 124 (2003) 1488–1499. doi:10.1016/S0016-5085(03)00276-2.

[170] H. Jaeschke, Emerging novel therapies against paracetamol (acetaminophen) hepatotoxicity, EBioMedicine. 46 (2019) 9–10. doi:10.1016/j.ebiom.2019.07.054.

[171] A. Kumari, S.K. Yadav, Y.B. Pakade, B. Singh, S.C. Yadav, Development of biodegradable nanoparticles for delivery of quercetin, Colloids Surfaces B Biointerfaces. 80 (2010) 184–192. doi:10.1016/j.colsurfb.2010.06.002.

[172] K. Hu, L. Miao, T.J. Goodwin, J. Li, Q. Liu, L. Huang, Quercetin Remodels the Tumor Microenvironment To Improve the Permeation, Retention, and Antitumor Effects of Nanoparticles, ACS Nano. 11 (2017) 4916–4925. doi:10.1021/acsnano.7b01522.

[173] P. Roy, S. Das, T. Bera, S. Mondol, A. Mukherjee, Roy, Das, Bera, Mondol, Andrographolide nanoparticles in leishmaniasis: Characterization and in vitro evaluations, Int. J. Nanomedicine. 5 (2010) 1113. doi:10.2147/IJN.S14787.

[174] S.M. Nabavi, S.F. Nabavi, S. Eslami, A.H. Moghaddam, In vivo protective effects of quercetin against sodium fluoride-induced oxidative stress in the hepatic tissue, Food Chem. 132 (2012) 931–935. doi:10.1016/j.foodchem.2011.11.070.

[175] M.I. Yousef, S.A.M. Omar, M.I. El-Guendi, L.A. Abdelmegid, Potential protective effects of quercetin and curcumin on paracetamol-induced histological changes, oxidative stress, impaired liver and kidney functions and haematotoxicity in rat, Food Chem. Toxicol. 48 (2010) 3246–3261. doi:10.1016/j.fct.2010.08.034.

[176] A. Budhian, S.J. Siegel, K.I. Winey, Haloperidol-loaded PLGA nanoparticles: Systematic study of particle size and drug content, Int. J. Pharm. 336 (2007) 367–375.

doi:10.1016/j.ijpharm.2006.11.061.

[177] A. Basu, S. Kundu, C. Basu, S.K. Ghosh, R. Sur, A. Mukherjee, Biopolymer nanoparticle surface chemistry dictates the nature and extent of protein hard corona, J. Mol. Liq. 282 (2019) 169–176. doi:10.1016/j.molliq.2019.03.016.

[178] K. Srinivas, J.W. King, L.R. Howard, J.K. Monrad, Solubility and solution thermodynamic properties of quercetin and quercetin dihydrate in subcritical water, J. Food Eng. 100 (2010) 208–218. doi:10.1016/j.jfoodeng.2010.04.001.

[179] P.A. McCormick, D. Treanor, G. McCormack, M. Farrell, Early death from paracetamol (acetaminophen) induced fulminant hepatic failure without cerebral oedema, J. Hepatol. 39 (2003) 547–551. doi:10.1016/S0168-8278(03)00299-X.

[180] D. Beales, A.M. McLean, Cell injury and protection in long-term incubation of liver slices after in vivo initiation with paracetamol: cell injury after in vivo initiation with paracetamol, Toxicology. 103 (1995) 113–119. doi:10.1016/0300-483X(95)03108-R.

[181] H.J. Zhao, M.J. Li, M.P. Zhang, M.K. Wei, L.P. Shen, M. Jiang, T. Zeng, Allyl methyl trisulfide protected against acetaminophen (paracetamol)-induced hepatotoxicity by suppressing CYP2E1 and activating Nrf2 in mouse liver, Food Funct. 10 (2019) 2244–2253. doi:10.1039/c9fo00170k.

[182] R. Tittarelli, M. Pellegrini, M.G. Scarpellini, E. Marinelli, V. Bruti, N.M. di Luca, F.P. Busardò, S. Zaami, Hepatotoxicity of paracetamol and related fatalities., Eur. Rev. Med. Pharmacol. Sci. 21 (2017) 95–101. http://www.ncbi.nlm.nih.gov/pubmed/28379590.

[183] L. Delahaye, E. Dhont, P. De Cock, P. De Paepe, C.P. Stove, Dried blood microsamples: Suitable as an alternative matrix for the quantification of paracetamol-protein adducts?, Toxicol. Lett. 324 (2020) 65–74. doi:10.1016/j.toxlet.2020.02.001.

[184] M. Naldi, M. Baldassarre, M. Domenicali, M. Bartolini, P. Caraceni, Structural and functional integrity of human serum albumin: Analytical approaches and clinical relevance in patients with liver cirrhosis, J. Pharm. Biomed. Anal. 144 (2017) 138–153. doi:10.1016/j.jpba.2017.04.023.

[185] J.R. Carvalho, M.V. Machado, New Insights About Albumin and Liver Disease, Ann. Hepatol. 17 (2018) 547–560. doi:10.5604/01.3001.0012.0916.

[186] M.K. Nguyen, I. Kurtz, Determinants of plasma water sodium concentration as reflected in the Edelman equation: role of osmotic and Gibbs-Donnan equilibrium, Am. J. Physiol. Physiol. 286 (2004) F828–F837. doi:10.1152/ajprenal.00393.2003.

[187] G. Feldmann, J. Penaud-Laurencin, J. Crassous, J.P. Benhamou, Albumin Synthesis by Human Liver Cells: Its Morphological Demonstration, Gastroenterology. 63 (1972) 1036–1048. doi:10.1016/S0016-5085(19)33181-6.

[188] D. Samuel, P. Ichai, Prognosis indicator in acute liver failure: Is there a place for cell death markers?, J. Hepatol. 53 (2010) 593–595. doi:10.1016/j.jhep.2010.06.002.

[189] J.A. Hinson, S.L. Pike, N.R. Pumford, P.R. Mayeux, Nitrotyrosine–Protein Adducts in Hepatic Centrilobular Areas following Toxic Doses of Acetaminophen in Mice, Chem. Res. Toxicol. 11 (1998) 604–607. doi:10.1021/tx9800349.

[190] S.K. Ramaiah, A toxicologist guide to the diagnostic interpretation of hepatic biochemical parameters, Food Chem. Toxicol. 45 (2007) 1551–1557. doi:10.1016/j.fct.2007.06.007.

[191] A. Eftekhari, E. Ahmadian, V. Panahi-Azar, H. Hosseini, M. Tabibiazar, S. Maleki Dizaj, Hepatoprotective and free radical scavenging actions of quercetin nanoparticles on aflatoxin B1-induced liver damage: in vitro / in vivo studies, Artif. Cells, Nanomedicine, Biotechnol. 46 (2018) 411–420. doi:10.1080/21691401.2017.1315427.

[192] A.H. GILANI, K.H. JANBAZ, B.H. SHAH, Quercetin exhibits hepatoprotective activity in rats, Biochem. Soc. Trans. 25 (1997) S619–S619. doi:10.1042/bst025s619.

[193] P.C.H. Hollman, M. V.D. Gaag, M.J.B. Mengelers, J.M.P. Van Trijp, J.H.M. De Vries, M.B. Katan, Absorption and disposition kinetics of the dietary antioxidant quercetin in man, Free Radic. Biol. Med. 21 (1996) 703–707. doi:10.1016/0891-5849(96)00129-3.

[194] H.A. El-Bakry, G. El-Sherif, R.M. Rostom, Therapeutic dose of green tea extract provokes liver damage and exacerbates paracetamol-induced hepatotoxicity in rats through oxidative stress and caspase 3-dependent apoptosis, Biomed. Pharmacother. 96 (2017) 798–811. doi:10.1016/j.biopha.2017.10.055.

[195] R. Domitrović, H. Jakovac, V. Vasiljev Marchesi, S. Vladimir-Knežević, O. Cvijanović, Ž. Tadić, Ž. Romić, D. Rahelić, Differential hepatoprotective mechanisms of rutin and quercetin in CCl4-intoxicated BALB/cN mice, Acta Pharmacol. Sin. 33

(2012) 1260–1270. doi:10.1038/aps.2012.62.

[196] K.. Janbaz, S.. Saeed, A.. Gilani, Studies on the protective effects of caffeic acid and quercetin on chemical-induced hepatotoxicity in rodents, Phytomedicine. 11 (2004) 424–430. doi:10.1016/j.phymed.2003.05.002.

[197] F. Sodano, L. Lazzarato, B. Rolando, F. Spyrakis, C. De Caro, S. Magliocca, D. Marabello, K. Chegaev, E. Gazzano, C. Riganti, A. Calignano, R. Russo, M.G. Rimoli, Paracetamol–Galactose Conjugate: A Novel Prodrug for an Old Analgesic Drug, Mol. Pharm. 16 (2019) 4181–4189. doi:10.1021/acs.molpharmaceut.9b00508.

[198] V.N. Uversky, Dancing Protein Clouds: The Strange Biology and Chaotic Physics of Intrinsically Disordered Proteins, J. Biol. Chem. 291 (2016) 6681–6688. doi:10.1074/jbc.R115.685859.

[199] M.M. Haque, R. Bayford, Protein Misfolding Thermodynamics, J. Phys. Chem. Lett. 10 (2019) 2506–2507. doi:10.1021/acs.jpclett.9b00852.

[200] J.T. Jarrett, P.T. Lansbury, Seeding "one-dimensional crystallization" of amyloid: A pathogenic mechanism in Alzheimer's disease and scrapie?, Cell. 73 (1993) 1055–1058. doi:10.1016/0092-8674(93)90635-4.

[201] Q.T.N. Tran, W.S.F. Wong, C.L.L. Chai, Labdane diterpenoids as potential anti-inflammatory agents, Pharmacol. Res. 124 (2017) 43–63. doi:https://doi.org/10.1016/j.phrs.2017.07.019.

[202] L. Wang, A. Tan, X. An, Y. Xia, Y. Xie, Quercetin Dihydrate inhibition of cardiac fibrosis induced by angiotensin II in vivo and in vitro, Biomed. Pharmacother. 127 (2020) 110205. doi:10.1016/j.biopha.2020.110205.

[203] T. Liu, Q. Yang, X. Zhang, R. Qin, W. Shan, H. Zhang, X. Chen, Quercetin alleviates kidney fibrosis by reducing renal tubular epithelial cell senescence through the SIRT1/PINK1/mitophagy axis, Life Sci. 257 (2020) 118116. doi:10.1016/j.lfs.2020.118116.

[204] Y.D. Álvarez, J.A. Fauerbach, J. V Pellegrotti, T.M. Jovin, E.A. Jares-Erijman, F.D. Stefani, Influence of Gold Nanoparticles on the Kinetics of α-Synuclein Aggregation, Nano Lett. 13 (2013) 6156–6163. doi:10.1021/nl403490e.

[205] H. Uchiyama, Y. Wada, Y. Hatanaka, Y. Hirata, M. Taniguchi, K. Kadota, Y. Tozuka,

Solubility and Permeability Improvement of Quercetin by an Interaction Between α-Glucosyl Stevia Nanoaggregates and Hydrophilic Polymer, J. Pharm. Sci. 108 (2019) 2033–2040. doi:10.1016/j.xphs.2019.01.007.